BIBLIOMETRIC STUDIES

BIBLIOMETRIC STUDIES

Editor
Dr. G. Devarajan
Deptt. of Library and Information Science
University of Kerala
Thiruvananthapuram

Ess Ess Publications
4837/24, Ansari Road, Daryaganj
New Delhi-110002
Phone: 3260807

CITS INFORMATION SCIENCE SERIES I

@ Centre for Information and Technology studies (CITS)
First Published 1997

ISBN : 81-7000-215-X

Price : Rs. Ess Ess Rs.700/-

Published by :

Ess Ess Publications

4837/24, Ansari Road, Darya Ganj,
New Delhi - 110 002
Phone : 3260807

Laser Typesetting :
CINOGRAPHICS
E-41, New Multan Nagar
New Delhi
Phone : 5583866

CONTENTS

PREFACE

Information managers have adopted a number of quantitative methods in recent years inorder to evaluate library resources and services more objectively and effectively. Bibliometrics is one of the quantitative techniques applied by information managers to measure the records of human communication. It is used to identify the pattern of publication, authorship, citations used for a subject etc over a period of time and thereby offering insight into the dynamics of the area under a particular study. Bibliometrics has attained significance in recent years because of its practical applications in various library operations and services. It is estimated that out of the total periodical literature published in Library and Information science at the global level, 25% ar on bibliometric studies. This data itself is a testimony to the growing importance of bibliometrics. Considering its contemporary significance an attempt is made by the Centre for Information and Technology Studies (CITS) to bring out a publication covering both the theoretical and practical aspects of Bibliometrics.

The Centre for Information and Technology Studies (CITS) Thiruvananthapuram is an organisation registered in 1993 under the Travancore- Cochin Charitable Societies Act, 1955 and its establishedment is the outcome of a long cherished desire of librarians in Kerala. It is a national level organisation and its Chairman is Prof. K.A. Isaac, Former Head, Department of Library and Information Science, University of Kerala. The following are the aims and objectives of the Centre.

i) to conduct studies in Information Science and Information Technology;

ii) to conduct courses in librarianship and Information Technology;

iii) to undertake continuing education programmes;

iv) to undertake projects and consultancy services;

v) to develop computer-based databases in Information Science and Information Technology;

vi) to establish a specialised library and documentation unit;

vii) to institute fellowships and awards; and

viii) to undertake publications in Information science and Information Technology.

Publication programme is one of the important objectives of CITS for dissemination of information in the field of Information Science and Information Technology. The Centre is publishing two series of books viz Books under Information Science series and Books under Information Technology series. The first book under Information Technology series entitled " Progress in Information Technology" had been released in January 1996. This book entitled "Bibliometric studies" is the first book under Information Science series and is the second book to be published by CITS. The Centre aims at bringing out one book every year under Information science series and Information Technology series.

There are 15 articles in this book each of which is dealing with a specific aspect of the broader subject area Bibliometrics. Most of the studies are the outcome of surveys and research conducted by authors in various fields. A select bibliography on Bibliometrics is included in this volume for the benefit of those who are interested in pursuing further study on bibliometrics. An alphabetical subject index is provided for the effective use of the text.

The Centre for Information and Technology Studies consider it a matter of pride that it was albe to publish two books in the field of library and Information Science. The editor on behalf of the Centre e x presses his indebtedness to the authors who contributed valuable articles to this volume. The Centre also expects the cooperation of the authors for the publication of other volumes of the series to be published in future.

The editor expresses his thanks to Ms. T. Ajikumary, Research scholar in Library Science who assisted in the compilation of the bibliography and Ms. Indu and Mr. Saji S Nair who assisted in the preparation of the index. Thanks are due to Ms. K. Litha who neatly typed the manuscript. The editor is also grateful to Mr. Sumit Sethi, M/S ESS ESS Publications, New Delhi who brought out this publication so neatly and in the best format.

Dr. G. Devarajan
(Editor and Secretary CITS)
Thiruvananthapuram
20.11.1996

CONTRIBUTORS

Accamma C Korah
Librarian
Rubber Research Institute of India
Kottayam, Kerala.

Dr. A. Amudhavalli
Reader
Department of Library and Information Science
University of Madras
Madras.

Aparna, P R
M.G. University
Kottayam.

Beena, S
Librarian
Dept. of Economics
University of Kerala
Thiruvananthapuram.

Chandrakumaran Nair, M
Librarian
Regional Cancer Centre
Thiruvananthapuram.

Dr. Humayoon Kabir, S
Lecturer
Dept. of Library and Information Science
M.G. University
Kottayam.

Dr. Jalaja, V
Lecturer
Dept. of Library and Information Science
University of Calicut
Calicut.

Jessey John
Kerala University Library
Thiruvananthapuram.

Dr. Maheswarappa, B.S.
Prof & Head
Dept. of Library and Information Science
Gulbarga University
Gulbarga.

Dr. Manorama Srinath
Reader
Dept. of Library and Information Science
University of Madras
Madras.

Mercy Jose
Documentation Officer
Rubber Research Institute of India
Kottayam.

Nargund, IN
Dept. of Library and Information Science
Karnataka University
Dharwad.

Rajamma, PK
Librarian
Central Tuber Crops Research Institute
Thiruvananthapuram

Dr. Sangam, SR
Reader
Dept of Library and Information Science
Karnataka University
Dharwad.

Sudhier, KG
VSSC Library
Valiamala
Thiruvananthapuram.

Usha Devi, SP
Asst. Librarian
Kerala University Library
Thiruvananthapuram.

Vijaykumar, KP
Lecturer
Dept. of Library and Information Science
University of Kerala
Thiruvananthapuram.

1

BIBLIOMETRICS : AN OVERVIEW

DR B S MAHESWARAPPA

0. INTRODUCTION

Bibliometrics is of recent origin and relatively a new one, which has emerged as a research front in its own right in Information science. It is now being vigorously pursued and with the result, it has been found that one-fourth of all the articles published in Library and Information Science periodicals are on bibliometrics and its related topics. It has also been found that many of the Social Science and Science periodicals are also carrying a large number of articles on bibliometrics. Pritchard and Witting[1] compiled a bibliography on the subject comprising 600 entries covering the period 1874 to 1959 which incidently rose to 2032 entries in 1980 as per compilation of Hjreppe.[2] In 1982, Hjreppe[3] again published a supplement to his 1980 bibliography covering 518 items of information. This increase in the number of publications may have been due to the use of bibliometric techniques in various disciplines and to the greater interest in the theory of literature structure. In view or the enormous literature available in the field- both in theory and applications, an attempt has been made in this article to provide an overview of bibliometrics, highlighting the original and important contributions to the field.

1. GENESIS OF BIBLIOMETRICS

Cole and Eale's[4] study on the "The History of Comparative Anatomy Part-I: A statistical Analysis" is considered to be the first bibliometric study, where for the first time in 1917, the expression `Statistical Analysis' has been used in the literature. Hulme[5] was the first to use the expression `Statistical bibliography' in 1923 and later it was used by many others. Gross and Gross's[6] study is considered to be the third study in the field based on citations. After Hulme, the term statistical bibliography was used by Henkle[7] in 1938 in his article "The periodical literature of Biochemistry" and Gosnell in his dissertation[8] in 1943, and later in his article[9] of 1984. Later Fusseler [10-11] in 1948 and 1949. Raising[12] in 1962, Barker[13] in 1966 and Pritchard[14-15] in 1968 and 1969 have used the term `Statistical Bibliography' in their works. The historical development of the term statistical bibliography, has been traced by Witting[16] in a foot-note. As the term was considered "very clumsy, not very descriptive, and can be confused with statistics itself or bibliographies on statistics''[17] Pritchard suggested the word 'Bibliometrics'[18] in 1969 in preference to statistical bibliography. Hence, the term bibliometrics has a very recent origin. The terms 'Librametrics', Scientometrics' and 'Informetrics' are also in use in the literature. Bibliometrics is analogous to 'Ranganathans' 'Librametrics' Russian concept 'Scientometrics', FID's 'Informetrics and also to some other well established subdisciplines like 'Econometrics', 'Psychometrics', 'Sociometrics' and 'Biometrics' where mathematical and statistical calculus have been systematically applied to study and solve the problems in the field of library science, history of science, information science, economics, psychology, sociology and biology respectively. As Kopelock[19] points out, during the period 1969-1977, altogether 52 different terms have been used to describe bibliometrics.

2. DEFINITION OF BIBLIOMETRICS

Many attempts have been made to define the term bibliometrics and its analogous terms since the use of the term 'statistical bibliography' in 1923 by Hulme. According to him,[20] the purpose of statistical bibliography is to "shed light on the process of written communication and of the nature and course of development of a discipline (in so far as this is displayed through written communication), by means of counting and analysis its various facets of written communication. "Raising[21] in 1962 defined it as "The assembling and interpretation of statistics relating to books and periodicals..... to demonstrate historical movements, to determine national and universal research, use of books and journals and to ascertain in many local situation the general use of books and journals." It is regarded as one of the classical definitions ofbibliometrics. As it was pointed out earlier, the term bibliometrics was first coined by Pritchard in 1969 in preference to existing terminology 'statistical bibliography' as he felt there is fair likelihood to misinterpret it as bibliography of statistics. He defined 'bibliometrics' as "The application of mathematical methods to books and other media of communication"[22]. According to Fairthorne,[23] it is the "Quantitative treatment of properties of recorded discourse and behaviour appertaining to it". The British Standard Glossary of documentation of Terms[24] explained bibliometrics as the study of "the use of documents and patterns of publication in which mathematical and statistical methods have been applied", which is basically similar to Pritchard's original definition. Hawkins[25] in his on-line bibliometric study interpreted bibliometrics as "the quantitative analysis of the bibliographic features of a body of literature." Nicholas and Ritchie[26] in their book entitled 'Literature on Bibliometircs' opined that bibliometrics....... provide information about the structure of knowledge and how it is communicated?". They further added that bibilometric studies fall mainly into two broad groups - those describing characteristics or features of a literature (descriptive studies) and those examining the relationship formed between the components of a literature (behavioural studies). More recently Potter[27], defined bibliometrics as "the study and measurement of the publication patterns of all forms of written communication and their authorship". Schrader[28] said it even more simply bibliometrics is "the scientific study of recorded discourse". Broadus[29] presented a historical overview of various definitions of bibliometrics and proposed an alternative definition. According to him, bibliometrics is the quantitative study of physical published units or of bibliographic units or of surrogates of either. More explicitly, Sengupta[30] defines it as the "Organisation, classification and quantitative evaluation of publication patterns of all macro-and micro communications along with their authorships by mathematical and statistical calculus".

The term 'Scientometrics' often used with the same meaning as the bibliometrics, originated in Russia to mean "the application of quantitative methods to the history of science,[31] but is now generally used as a generic term for a variety of research approaches within the study of science.... that quantifiable aspects of science can be utilized to assess the characteristics of science. Merton and Garfield[32] have defined it as the field of enquiry given over to the quantitative analysis of science and scientific field... Come to be known as Scientometrics or bibliometrics. According to Sen[33] bibliometrics deals with documents and its components while metric studies pertaining to information is informetrics. Morales[34] uses the term informetrics to cover almost all the aspects of bibliometrics and librametrics. All those four metrics have overlapping areas.

3. BIBLIOMETRIC LAWS

The three fundamental laws which laid the formation of bibliometrics are :

(1) Lotka's Inverse Square Law of Scientific Productivity:

(2) Bradford's Law of Scattering of Scientific Papers; and

(3) Zipf's Law of Word Occurence.

Lotka's Law:

In 1926, Alfred J. Lotka proposed his inverse Square Law correlating contributors of scientific papers to their number of contributions. His law provided fundamental theoretical base for bibliometric studies involving authorships. He was interested in determining "the part which men of different calibre contribute to the progress of science"[35]. For this, he checked the decennial index of 'Chemical Abstracts' 1907--1916 and counted the number of names against which appeared 1,2,3 etc., entries, He tabulated the data for 6,891 names, beginning with letter`A' and `B' Similarly the data from the Auerbach's Geschietftafeln der physik was also collected for the 1325 physicists. Lotka then plotted the graph on a logarithmic scale, the number of authors against the number of contributions made by each author and he found that in each case the points were closely scattered about a straight line, having a slope of approximately two to one. On the basis of these data, Lotka deduced a general equation, for the relation between the frequency `y' of persons making `x' contributions as follows:

$$x^{n}.y = \text{constant}$$

and for the special case n = 2, the constant is 0.6079. Further he summarised the results as follows :

"In the case examined it is found that the number of persons making 2 contributions is about one-fourth of those making one contribution, the number making `n' contributions is about $1/n^2$ of those making one and the proportion of all contributions is about 60 per cent."[36]

In other words, for every 100 authors contributing one article, 25 will contribute two articles, about 11 will contribute 3 articles and 6 will contribute 4 articles and so on. Though, the law was based on the study of chemistry and physics literature later it has generated much interest and attracted the attention of researchers and it has been applied and tested in many other fields.

Bradford's Law of Scattering:

Samuel Clement Bradford,[37] another pioneer of bibliometrics, should be considered for his classic paper "Sources of Information on Specific Subjects," which is the first paper published on observations on scattering. Bradford examined two bibliographies prepared in the Science Library (Britain) on Applied Geophysics (1928-31) and Lubrication (1931-32) and he prepared lists of journals arranged by decreasing order of source items contributed by the journals to the bibliographies. He noticed that in each subject there were a few very productive sources, large number of sources which were moderately productive and still a large number of sources of constantly diminishing productivity. The whole range of periodicals was thus seen as "a family of successive generations of diminishing kinship, each generation being greater in number than the preceding, and each constituent of generation producing inversely according to its degree of remoteness."[38]

In the list of periodicals ranked by diminishing productivity, Bradford identified three groups of periodicals that produced approximately the same number of articles on the subject, but the number of periodicals in these three equiproductive zones increased by a constant factor. Based on this he stated his law as follows : "If scientific periodicals are arranged in order of decreasing productivity of articles on a given subject that may be divided into a nucleus of periodicals more particularly devoted to the subject and several groups or zones containing the same number of articles as the nucleus when the number of periodicals in the nucleus and succeeding zones will be as 1 : n : n^2"[39].

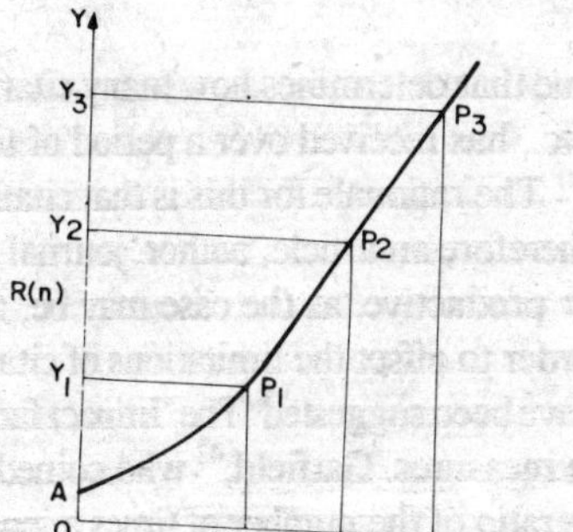

Fig. 1

Bradford also plotted graphs of the cumulative number of source items R(n) versus the logarithm of the cumulative number of journals (log n). The resulting graphs for Applied Geophysics and Lubrication were similar to the graph shown in Fig. 1. Such a graph, is sometimes called as Bradford bibliograph.

The graph begins as a rising curve API and then continues as a straightline. The rising part of the graph represents the nucleus of highly productive journals. The points p1, p2, and p3 on the bibliograph are the boundaries of three equiproductive zones in which the same number of articles as the nucleus (represented by OY1 = y1y2 = y2y3) derived from an increasingly larger number of journals (represented by OXI, X1, X2 and X2, X3). The law attracted the attention of many researchers in the field and has been the main topic of many articles in the literature.

Zipf's Law of Word Occurence:

Zipf[40] developed and extended an empirical Law, as observed by Estoup, governing a relation between the rank of a word and the frequency of its appearance in a long text. If 'r' is the rank of a word and 'f' is its frequency, then mathematically Zipf's law can be stated as follows :

rf = c, where 'c' is a constant.

His law states that in a long textual matter if words are arranged in their decreasing order of frequency, then the rank of any given word of the text will be inversely proportional to the frequency of occurance of the word. Thus, these three laws are respectively based on (i) number of authors contributing in a discipline or other field; (ii) distribution of articles in a set of journals; and (iii) ranking word frequency in a particular set of documents.

4. BIBLIOMETRIC MEASUREMENTS

The bibliometric measurements are derived from the concept of Citation indexing. It was based on the English Legal System, which operates under the doctrine of 'stare deisis precedent,' on the basis of which Garfield developed 'Science Citation Index,' 'Social Science Citation Index and Arts and Humanities Citation Index'.

Direct Citation Counting:

Citation counting is a technique that determines how many citations a given document, author, journal etc., has received over a period of time, originally used by Gross and Gross.[41] The rationale for this is that citations are objective indicators of use and therefore an article, author, journal that is frequently cited, is more useful or productive, as the case may be, than one that is less frequently cited. In order to offset the limitations of citation counting, some modified measures have been suggested. The 'Impact factor' and immediacy Index' are two such measures. Garfield,[42] who coined the term impact factor defines it as "the ratio of the number of times a journal is cited in a given time period to the total number of times a journal is cited in a given time period to the total number of source items published in the journal, during specified period of time". The impact factor is a measure of the frequency with which the average cited article in a journal has been cited in a particular year. It offsets the effect of age, size the frequency of a publication of a journal on the frequency of citation. The immediacy Index is a method of showing the frequency with which a material received by the article during the year to the number of articles published. Recently, Hirst[43] has suggested the discipline impact factor (DIF), a method of determining core journals for a discipline, which is similar to the impact factor, which measures the number of times a paper in a journal is cited in the core literature of the given discipline.

Bibliographic Coupling:

The concept of bibliographic coupling was first suggested by Fano,[44] but Kessler[45] elaborated, tested and coined the term. It is the number of common references cited in two document that indicates the degree of similarity of contents of the citing papers. Two source documents containing a large number of common references are said to have a high coupling strength and are likely to be on the same topic. It is observed that the concept of relationship has certain drawbacks and not seem to be a valid unit of measurement because if two papers are citing a third paper, they may or may not be citing an identical piece of information of third paper being cited.[46] Further Tagliacozzo[49] is of the opinion that "the fact that two papers have a reference in common is no guarantee that both papers are referring to the same piece of information. So, it is merely an indication of the existence of the probability of relation between two documents".

Co-citation:

The concept of Co-citation was for the first time suggested independently by Small[48] and Marshakova[49] almost simultaneously in 1973 and later developed by Small, who proposed a new method of analysing citations, to generate clusters of related papers. The number of times two papers are cited together in subsequent literature determines the co-citation strength of two cited papers. Co-citation is dynamic measure in that co-citation strength of cited papers can be studied over a period of time as they continue to be cited together in subsequent literature. But, one of the disadvantages of cocitation technique is that, it requires comprehensive citation data.

5. BIBLIOMETRIC MODELS

The empirical laws of bibliometrics have been discussed earlier in this paper. In order to provide a sound conceptual understanding of these laws, many bibliometric models have been developed so far. Some of the important bibliometric models are : Logarithmic Model of Brookes;[50] Cumulative Advantage Distribution Model by Price;[51] Bookstein's Model[52] linking laws of Bradford, Lotka and Zipf into a single law that seems capable of describing phenomenon in a vast variety of subject areas; Mandelbrots[53] Model structure of language expending and restating Zipf's law; Fairthorness[54] Model linking Lotka's law with the distribution of Bradford, Zipf and Mandelbrot. Vickery's[55] Model extending Bradford's distribution and showing that the relation should hold for any number of zones of equal yields, not just for high, medium and low yielding zones as proposed by Bradford; graphical models of Leimkuhler,[56] Goffman and Warren,[57] Sengupta's Model[58] providing an of setting weightage formula for postwar scientific periodicals; Hill''s[59] Model negative binomial distribution; Naranan's [60] power law model and many others.

6. BIBLIOMETRICS AND ITS APPLICATIONS

Bibliometrics as a technique has extensive applications in identifying the research trends in a subject, trends in authorship and collaboration in research; core periodicals; obsolescence, and dispersion of scientific literature useful in estimating the comprehensiveness of secondary periodicals, studying the author productivity; characteristics of subject literatures including structure of knowledge, historical and sociological aspects of science and helpful in formulation of need based collection development policy, weeding and stacking policy, science policy studies and many others.

REFERENCES:

1. Pritchard, Alan and Witting, G R : Bibliometrics: A bibliography and index (1874-1959). Wettford, AAIHBOOKS, 1960.
2. Hjreppe, R: A Bibliography of bibliometrics and citation indexing and analysis. Stockholm, Royal Institute Technology Library, 1980.
3. Hjreppe, R : Supplement to a Bibliography of bibliometrics and citation indexing. Scientometrics 4 : 1982, pp. 241-273.
4. Cole, F J and Eales, N B : The History of Comparative Anatomy part 1. - A Statistical analysis of Literature Science Progress, 11: April 1917, pp. 578-96.
5. Hulme, E Wyndham : Statistical bibliography in relation to the growth of Modern Civilization. London, Butler and Tanner Grafton, 1923. p.9.
6. Gross, PLK and Gross F M : College Libraries and chemical Education. Science, 66: 1927. pp. 386--389.
7. Henkle. H H : The periodical Literature of Biochemistry. Bulletin of the Medical Library Association, 27: 1938, pp. 139-147.
8. Gosnell, Chas F : The rate of obsolescence in college library Book collections as determined by the analysis of Three select list of Books for College Libraries. (Dissertation; New York University, 1943).
9. Gosnell, Chas F : Obsolescence of Books in College Libraries. College and Research Libraries, 5: 1944, pp. 115-125.
10. Fusseler, Herman H : Characteristics of the Research Literature used by chemists and physicists in the United States (Dissertation; University of Chicago, 1948).
11. Fussler, Herman H : Characteristics of the Research literature used by chemists and physicists in the United States. Library Quarterly, 19: 1949, pp. 19-35.
12. Raising, L Miks : Statistical Bibliography in the Health Sciences. Bulletin of the Medical Library Association, 50: 1962, pp. 450-461.
13. Barker, Dale Lockard : Characteristics of the scientific Literature cited by chemists of the Soviet Union (Dissertation; University of Illinois, 1966).
14. Pritchard, Alan : Computers, statistical bibliography and Abstracting Services, 1968 (unpublished).
15. Pritchard, Alan : Statistical Bibliography : An interim bibliography. London, New Western Polytechnic School of Librarianship, 1969.
16. Witting Glenn R : Statistical Bibliography: A Historical Footnote. Journal of Documentation, 34: 1978. pp. 240-241.
17. Pritchard, Alan : Statistical Bibliography or Bibliometrics. Journal of Documentation, 25(4) : 1989, p.348.
18. Ibid, p. 348.
19. Kopelock, B : Bibliometrics : Access in the library literature. Collection Management, 2(3) : 1978. pp. 199-204.
20. Hulme, F W : op.cit, p.21
21. Raising, L M : op.cit pp. 450-461
22. Pritchard, Alan : op.cit p. 349
23. Fairthorne, R A : Empirical hyperbolic distributions (Bradford. Zipf-Mandelbort) for

bibliometric description and prediction. Journal of Documentation, 25 : 1969. p. 319.

24. British Standards Institution : British Standards of Documentation Terms. London; BSI, 1976. p. 7.

25. Hawkins, D T : Unconventional uses of on-line information retrieval systems: on-line bibliometric studies, Journal of American Society for Information Science, 28(1) : 1977, pp. 13-18.

26. Nicholas, D and Ritchie, M : Literature on Bibliometrics. London, Bingley, 1978, pp. 9-10.

27. Potter, W G : Introduction to Bibliometrics. Library Trends, 30 : 1981. p. 5.

28. Schrader, Alvin M : Teaching Bibliometrics. Library Trends. 30. 1981. p. 151.

29. Broadus, R N : Definition of bibliometrics, Scientometrics 12 : 1987. pp. 373-398.

30. Sengupta, I N : Bibliometrics and its applications .In : Information Science and Libraries ed. by Pushpa Dhyani. New Delhi, Atlantic, 1990. p. 256.

31. Dobrov, G H and Korennoi, A A : The Informational basis of scientometrics : on theoretical problems of informetrics. F. I. D, 435 1969, pp. 165-191.

32. Merton, R K and Garfield, E : Forword to little science, big science - and beyond by DJ de Solla Price. New York Columbia University Press, 1963.

33. Sen, SEN : From Bibliometric to informetrics: Workshop on scientific communication and Bibliometrics held at ICCB. and INSDOC, Calcutta, Feb. 15-28, 1988 pp. 84-98, (preprint).

34. Morales, M : Informetrics and its importance, International Forum for Information and Documentation, 10 (2) : 1985. pp. 15-21.

35. Lotka, A J : The frequency distribution of scientific productivity, Journal of Washington Academy of Sciences, 16: 1926. p. 317.

36. Ibid, p. 323.

37. Bradford, S C : Sources of Information of specific subjects. Engineering, 137 : 1934. pp. 85-86.

38. Ibid.

39. Ibid : p. 86.

40. Zipf, G K : Human Behaviour and the principle of least effort : An introduction to human ecology. Mass Addison-Wesely, 1949.

41. Gross, P L K and Gross, E M : op. cit. pp. 384-389.

42. Garfield, Eugene : Citations Analysis as a tool in journal evaluation, Science 178 : 1972. pp. 471-478.

43. Hirst. Grame : Discipline Impact Factors : A method for determining core journal lists. Journal of the American Society for Information Science, 29(4) : 1978. pp. 171-172.

44. Fano, R M In : Documentation in Action New York, Rein-hold 1956. pp. 238-244.

45. Kessler, M M : Bibliographic coupling between scientific papers. American Documentation, 14 : 1983. pp. 10-15.

46. Martyn, J : Bibliographic coupling. Journal of Documentation, 20 : 1964. pp. 236.

47. Tagliacozzo. Citations and Citation Indexes. Match. Inform. Med, 6 : 1967 . pp. 136-142.

48. Small, Henry : Co-citation in the Scientific Literature: A new measure of relationship

between two documents. Journal of American Society for Information Science, 24 (4) : 1937. pp. 265-69.

49. Marshakov, I V : System of document connection based on references. Nauchne-Tekumicheskala Informatsija, 1973 series 2. pp. 3-8.

50. Brookes, B C : Theory of Bradford's Law, Journal of Documentation, 33 (3) : 1977. pp. 180-209.

51. Solla Price, D J de : A general theory of Bibliometrics and other cumulative advantage process. Journal of American Society for Information Science, 27 (5/6) : 1976. pp. 292-306.

52. Bookstein, A : Explanation of bibliometric laws. Collection Managment 3 : 1979. pp. 151-162.

53. Mandelbort : An Information theory of statistical structure of language. In : Proceeding of symposium on application of communication theory. London, Butterworths, 1953. pp. 486-500.

54. Fairthorne, R A : op.cit. pp. 319-343.

55. Vickery, B C : Bradford's law of scattering. Journal of Documentation, 4 (3) 1948. pp. 198-203.

56. Leimkuhler, F F : Bradford distribution, Journal of Documentation, 23 (3) : 1967. pp. 197-207.

57. Goffman, W and Warren, K S : Distribution of papers among journals based on mathematical analysis of two diverse medical literature. Nature, 221 : 1969. pp. 1205-1207.

58. Sengupta, I N : A weightage formula to remark biochemical periodicals. International Library Review, 16 : 1984 pp. 447-458.

59. Hill, B M : Zipf's law and prior distribution for the composition of a population. Journal of American Statistical Association, 65 : 1970. pp. 1220-1232.

60. Naranan, S : Power law relations in science : A self consistent interpretation. Journal of Documentation, 29 (2): 1971. pp. 834-97.

2

SCIENTOMETRICS/INFORMETRICS: EVOLUTION OF ITS CONCEPT AND APPLICATIONS

DR. A. AMUDHAVALLI

INTRODUCTION

In the U.S., the beginnings of Information Science' were marked in the 1950s, by the migration of a number of natural scientists, behavioural scientists, computer scientists and applied mathematicians into problem areas in the information field. In the recent past, the bulk of the contributions influencing the body of 'Information Science' seem to have stemmed from mathematical and statistical sciences. Perhaps the founders of 'Information Science' have applied the name in the belief that they are talking of a body of knowledge capable, in some degree, of quantitative analysis. "The emphasis on the word Science' reflects not only a desire for professional status but also the hope that it will be possible to develop the solid underpinnings of theory and tested factual characteristics of the other sciences". One of the basic and uncontested (also unproved) beliefs in science studies is the assumption that scientific progress is a cumulative (and hence measurable) growth of knowledge.

The knowledge of scientific literature has been augmented by a considerable body of statistical research. This has been directed towards various phases of scientific communication, but has sought principally to describe the characteristics of the literature available to scientists to identify the literature required by them, or to determine how they use the literature.

The clear truth is that the phenomena of exponential growth of scientific literature on 'Information Science', coupled with the availability of sophisticated computer based statistical software packages for data analysis have encouraged quantitative and qualitative studies. The outcome of such efforts is the emergence of the new science, whose prime emphasis is the application of the mathematical and statistical models and methods to study and measure the media of communication, particularly graphic media. Analysis of published literature can be done by numerous methods. Broadly, they can be grouped into qualitative and quantitative methods. The qualitative methods are based on application of value systems and standards as found in 'Humanities'. The quantitative methods, on the other hand, are based on ''application of numerical and experimental methods as found in the 'Sciences'. The quantitative method provides critical comparisons but its weakness lies in possible subjectivity and incompleteness of the overall picture. The quantitative method has the ability to cover regularities and patterns in behaviour, together with the disadvantage of dealing only with surface and not with underlying causes and qualities. Of all the bibliographic characteristics, subject/content of the literature is the most significant element of quantitative studies. As a majority of bibliometric studies were conducted within a subject framework, it is also of paramount interest to study the relationship of the host subject to other subjects. This kind of information on subject relationships may lead to the identification and ranking of subjects contributing to the given subject (the extent to which it draws from the literature of other sciences). Thus both composition and distribution imply inter-relationships. However, measuring knowledge is always a difficult task, as bibliographic descriptions seldom go further than mapping the exterior features of documents and their value relies and depends on the extent to which the document's outward features divulge the inner self, though the contents of a document is what is of real importance.

This science is being referred to as 'Bibliometrics/Scientometrics/Informetrics'. (There are differences regarding the connotations of these terms. These, however, are not relevant to this study and hence these differences will not be examined here. The three terms will be used more or less synonymously in this study.) This subject has gained significance for the simple reason that it is not merely a theoretical discipline, but has extensive applications. It is used to identify the pattern of publication, authorship, citations and behaviour of a subject over a period of time or at a given time and thereby offering insight into the dynamics of the area under study. The recent area of interest is what it offers to the sociologist and scientist of knowledge. This area of study, emerging as a principal contributor to the development of 'Information Science' as an interdisciplinary field, warrants an investigation as to the evolution of its concept, meanings and applications.

The desire to make evaluative and quantitative statements about information is another themes evident from early days to modern 'Information Science'. This impetus may have come partly from Shannon's 'Information Theory', (the origin of its concept even a good deal earlier) and partly from the development of the new scientific discipline, 'Science of Science' (the origin of which is only a recent phenomena). Its emergence is associated with Bernal's book. According to him, the 'Science of Science' (consciousness of science) has been a great undertaking of the second half of the twentieth century.

The pre-1950s literature has evidences of the key contributions to the fundamental concept, and coinage of the term 'Statistical Bibliography', that later changed to 'Bibliometrics'. The classical works of Pritchard, Fairthorne, Lotka, Bradford, Zipf are worth mentioning in this context.

The post-1950s work in the field has seen the emergence of a 'Science of Science', largely based on contributions of Derek Price and the citation indexes of Eugene Garfield. The works of Price on the qualification of science and in particular of scientific output and communication and the application of mathematical statistics for analyzing the scientific output and the process of scientific development had a notable effect on the studies in this field. It was about then that related studies. These works stimulated and commanded a growing perception of the need for objective measures of scientific productivity amongst numerous researchers world-wide resulting in a fleet of research contributions to this area employing simple to sophisticated statistical techniques. Its accelerated development began in the '60s, when a hindrance effect of a number of parameters characterizing the state of science became apparent, and in the course of this development several terms including Statistical Bibliography, Librametry, Bibliometrics, Scientometrics and Informetrics were brought into usage. The evolution, scope and definitions of these terms go to emphasize the progress and the significance of quantitative techniques and their contribution to the field of 'Information Science'.

Two terms used interchangeably and most popularly used currently are 'Scientometrics' and 'Informetrics', which are discussed subsequently.

1.SCIENTOMETRICS

This term was introduced and came into prominence with the founding of the journal named 'Scientometrics' by T. Braunin 1977, originally published in Hungary and currently from Amsterdam. The scope of the journal is to publish all those studies of quantitative aspects of science as a discipline or economic activity. Scientometrics is part of the sociology of science and has application to science-policy making. It involves quantitative studies of scientific activities, including among others, publication, and so overlaps bibliometrics to some extent. Scientometrics is branch of the 'Science of Science'. Nalimov and Mulchenko define this term, "as a sub-field which applies quantitative methods to the study of science as an information process". In this information model, publications are carriers of information and promoters of communication. Haitun treats 'Scientometrics', as a scientific discipline which performs reproducible measurements of scientific activity and reveals its objective quantitative regularities. According to him, scientometric methods include statistical and thesaurus methods, and indicators as to the number of citations, terms etc. There are two aspects within science of science, viz.:

1. the analytical aspect which deals with the general laws of the development of science as a knowledge system and a specific social institution; and
2. the normative aspect which deals with the development of practical recommendations for raising research efficiency.

The principal aim of 'Scientometrics' is to determine the state and prospect of a subject and its further development. Several scientometric indicators are used for this purpose, and one of the most significant indicators is the number of publications; its changes over time is usually considered as a measure of research topicality in a given field. These indicators are on the way to become a standard tools of evaluation and analysis in research management of science policy making. Considerable attention has been paid in recent years to ways to measure scientific information, which may be used for the control of science development in two ways: practical and strategic. The practical way includes search of information to users. As far as the strategic use of information is concerned it implies the monitoring of science development and the analysis of the changing structure of science.

Since the beginning of 1970s, this kind of research activity aimed at getting structural (or qualitative) picture of the state of science is gaining popularity.

The mathematical-statistical techniques had made possible the process of quantifying the number of publication and the number of people participating in the social system of science, research or R & D. Although there might be methodological difficulties, still it was relatively easy since it was (or can be reduced to) a process of counting. The question, however, is whether such a procedure is really adequate if one has to do with the quantification of the 'intellectual products' of scientists or researchers. Can the information contained in the graphic media be measured in the same way as the other one? This presupposes the emphasis on qualitative studies which should be descriptive in nature. This notion has prompted researchers in this field, in the decades, to develop special type of procedures and quantitative techniques to study the features of a given discipline and to observe changing structural pattern of the discipline over a period of time. For example, proponents had pointed to:

* an ability to map literature of a discipline intelligibly at different levels;
* an ability to create useful indicators of developments in science and increasingly in technology;
* an ability to model dynamic aspects of literature mathematically with good-to-excellent fits;
* growing skill in using computerized databases and practical software for bibliometric ends.

Research along these lines were particularly evident in the literature output from France (Michael Callon and others) Netherlands (A F J Van Raan), USSR (A P Trofimenko) etc.

2. INFORMETRICS

According to Brookes, the fourth term 'Informetrics' was first proposed by Otto Nacke of West Germany in 1979[6]. An FID Committee with very broadly defined objectives in the provision of research and technical data was subsequently given this name. However, the term was not widely adopted until 1987 when B.C. Brookes at the First Conference on Bibliometrics & Theoretical Aspects of Information Retrieval held at Diepenbeek Belgium, suggested that the term 'Informetrics" be included in the name of the Second International Conference on the subject. Egghe and Rousseau in the proceedings of the conference state that "the term 'Informetrics' was favoured rather that the terms (or beside) the terms 'Bibliometrics' and 'Scientometrics'. Hence, in promoting a new name, it was decided to use the name 'Informetrics' together with the name 'Bibliometrics in the title of the following conferences and in the title of the published conference proceedings also"[17]. The second meeting was thus named as 'International Conference on Bibliometrics, Informetrics and Scientometrics' and the term 'Informetrics' continues to be used in this series of biennial conference.

Brookes opines that "this new term is being used to cover both sciento- and biblio- metrics impartially. It has produced no distinctively new ideas of its own but as it implicity covers both documentary and electronic forms of communication, it may have a future"[5] As a general field of study, it includes the earlier fields `Bibliometrics' and `Scientometrics'.

Tague-Sutcliffe defines the term, as "the study of the quantitative aspects of information in any form, not just records or bibliographies, and in any social group, not just scientists........ It can incorporate, utilize, and extend the many studies of the measurement of information that lie outside the boundaries of both `Bibliometrics' and `Scientometrics', and she continues to say that, "Although in practice the scope of `Informetrics' is very broad, ........... two phenomena that have not, in the past, been seen as part of `Bibliometrics' and `Scientometrics', but fit within the scope of `Informetrics' are:

* definition and measurement of information, and
* types & characteristics of retrieval performance measures".

The former has been studied in 'Electrical Engineering' and 'Mathematics' since Shannon and Weaver (1949), and versions of the original measure have, indeed, been developed in many fields. Retrieval performance measures have been studied by information retrieval theorists in 'Information and Computer Sciences' for both its theoretical and practical aspects. Thus, it can be said that the scope of 'Informetrics' is both practical and theoretical. However, the primary emphasis has been on the development of mathematical models, and a secondary emphasis on the derivation of measures for the diverse phenomena studies. The value of a model lies in its ability to summarize, in terms of a few parameters, the characteristics of many data sets: the overall shape, concentration, scatter, and the way the data sets change over time. Such models provide a basis for practical decision-making.

This historical account confirms that the contemporary approaches in 'Scientometrics'/'Informetrics' are marked by extensive use of quantitative and qualitative techniques. In recent years, such quantitative methods are being extensively used for studying the structure of literature (of discipline). These techniques have been helpful in enabling the mapping of disciplines as also a study of any transition in the structure and composition of a discipline. In fact, in the last one decade a number of tools, techniques and indices for this purpose have been developed and applied to various areas.

REFERENCES:

1. Bradford, S C : Sources of information on specific subjects. Engineering, 137: 1934. pp. 85-86.
2. Garfield, Eugene : Citations analysis as a tool in journal evaluation. Science 178: 1972 pp. 471-478.
3. Lotka, A J: The frequency distribution of scientific productivity. Journal of Washington Academy of Sciences, 16: 1926.
4. Morales, M : Informetrics and its importance. International Forum for Information and Documentation. 10 (2): 1985 PP. 15-21.
5. Nicholas, D and Ritchie, M : Literature on bibliometrics. London, Bingley, 1978.
6. Potter, W G : Introduction to bibliometrics. Library trends. 30 : 1981.
7. Solla Price, D J de : A general theory of bibliometrics and other cumulative advantage process. Jl of American Society for Information Science 27 (5/6): 1976. pp. 292-306.

8. Tagliacozzo : Citations and Citation indexes. Match Inform. Med. 6:1976. pp. 136-142.

3

APPLICATIONS OF BIBLIOMETRICS IN LIBRARIES AND INFORMATION CENTRES

K.P. VIJAYAKUMAR

0. INTRODUCTION

Though the term bibliometrics was introduced only in 1969 to indicate a new discipline which employs quantitative methods for analysing various aspects of written documents its origin can be traced back to the efforts of early twentieth century documentalists to apply mathematical and statistical analysis to bibliographical units. The most prominent efforts were that of Cole and Eales[1], Hulme[2], Lotka[3], Zipf[4], Bradford[5], Ranganathan[6], Price[7], Kessler[8], Garfield[9], Goffman & Warren[10] and Pritchard[11]. The history of this discipline is discussed by Egghe[12].

Of the various prominent studies mentioned above, three laws became the cornerstones of bibliometrics. They were:

i. Lotka's inverse square law about the productivity of authors;

ii. Zipf's law about the frequency occurrence of words in a text; and

iii. Bradford's law of scattering of articles.

The paper by Maheswarappa in this volume deals with these laws in greater detail. Several attempts have been made to bring out the interrelationship among these three laws. Examples are Narin and Moll[13], Price[14], Meadows[15], Fairthorne[16], Bookstein[17], Chen and Leimkuhler[18]. In fact, much of the later bibliometric studies were made either to substantiate, modify, extend, link or challenge these three empirical laws. The end result of such studies was the emergence of some more empirical laws and many theoretical models. Some of the other empirical laws are:

i. Price's Square Root Law of Scientific Productivity

This law states that "half of the scientific papers are contributed by the square root of the total number of scientific authors[19]. In other words, $N^{1/2}$ sources yield a fraction 1/2 of the items. This phenomenon is associated with the occurrence of invisible colleges. This law is sometimes called `Rousseau's law' since Jean Jacques Rousseau had mentioned the same thing quite clearly in his `Social Contract' about the size of the elite, i.e. those participating in the government. Egghe and Rousseau argue that Price's law is not generally valid[20]. This can also be treated as an extension of the success-breeds-success principle originally developed by Simon in 1955.

ii. Garfield's Law of Concentration

Garfield talked about the number of journals involved in publishing the literature of a single field[21]. He did not say anything about how much the journals in one field might overlap with other fields. In fact, there is a significant degree of overlap. Several studies have shown that relatively few journals are involved in the publishing of an overwhelming majority of the material in a subject. A study of the Science Citation Index (SCI) database showed that 500 journals accounted for 70% of the material indexed in SCI in 1969. Almost half of the 3.85 million references published that year was found to emanate from only 250 journals. This type of evidence makes it possible to move from Bradford's law of dispersion to Garfield's law of concentration.

The law states that "a basic concentration of journals is the common core or nucleus of all fields". In other words, the tail of the literature of one discipline consists, in a large part, of the cores of the literature of other disciplines. So large is the overlap among disciplines that the core literature of all scientific disciplines involves a group of not more than 1000 journals.

iii. Sengupta's Law of Bibliometrics

This is basically an extension of the Bradford's law. It states that "during phases of rapid growth of knowledge in a scientific discipline, articles of interest to that discipline appear in increasing number of periodicals distant from that field"[22]. Mathematically Sengupta's law stands in the following form:

$$f(x+y) = a+b \log (x+y)$$

where $f(x+y)$ is the cumulative number of references as contained in the first $(x+y)$ most productive journals, x indicate number of journals in the same discipline and y stands for number of journals of unrelated disciplines $(y>x)$ and a and b are two constants.

Ravichandra Rao summarises other empirical laws in one of his papers and those who are interested can go through the reference. He has also listed more important bibliometric models.[23]

1. QUANTITATIVE AND QUALITATIVE STUDIES

The discovery of the empirical laws of bibliometrics has led to a series of studies which can be broadly differentiated into two--: quantitative and qualitative. In fact, the early statistical studies of Cole and Eales, Hulme, Lotka, Zipf and Bradford belong to the quantitative category. B.C. Brookes is of opinion that such quantitative studies have five general objectives:

i. Design of more economic information systems and networks;

ii. Improvement of efficiency rates of information handling process;

iii. Identification and measurement of deficiencies in bibliographical services;

iv. Prediction of publishing trends; and

v. Discovery and elucidation of empirical laws that can provide a basis for developing a theory of Information Science[24].

On the other hand, qualitative application studies emphasize practical utilization of research findings. Some of the findings which are fruitfully used in library management are:

i. Identification of a core literature;

ii. Ranking publications in zones of diminishing importance;

iii. Establishing a transition point between zones of higher and lower utility;

iv. Tracing the spread of ideas as a study of epidemics; and

v. Classifying segments of a literature through interconnection of citations[25].

Two more points are added by Susan Artandi. They are:

i. determination of the impact value of a given document;

ii. location of the criticisms of published results of research and experiments[26].

Some of the above points will be discussed while examining the application of citation analysis. Whatever may be the type of study, Burton is of opinion that the product of bibliometric analysis is metainformation or information about information[27]. This metainformation is of use to the following three categories of people:

i. To the librarian or information specialist as a means to better understand the environment in which service and support are offered;

ii. To the research manager as a means to measure and evaluate productivity of his or her own staff and how it compares to other comparable units; and

iii. To the research analyst who may wish to define a new research project or compare his or her own progress to that of a broader spectrum - such as other institutions, related disciplines, or other nationalities.

2. APPLICATIONS OF BIBLIOMETRIC LAWS

2.1 Bradford's Law

The statistical regularity pointed out by Bradford's law provides an objective means of determining zones of relative richness or value to a given kind of library collection. This has implications to the acquisition process in a library. A library can safely stock the journals which belong to the core or nuclear zone. It is advisable to extend the purchase list to the next zones till the budget limits permit. If at all the budget is elastic, a point will be reached at which it would be desirable to obtain copies of articles in the journals on demand rather than subscribing to the journal. Lancaster provides an excellent hypothetical example of applying Bradford's law in periodical collection building, while discussing the principle of diminishing returns[28]. Brookes is of the view that if the total expenditure on periodical provision is limited to the fraction 'f' of the sum needed to cover the subject completely, the buying of periodicals may be supplemented by the buying of photocopies of the relatively few relevant papers published in the peripheral periodicals.

While preparing bibliographies we are faced with the problems of coverage, the journals that are to be scanned etc. Bradford's distribution can be fruitfully used to estimate the total size of a bibliography and the periodicals that should necessarily be included in the list of items to be covered.

On the application side of this law, the studies of Goffman and Morris[29] and Ravichandra Rao[30] are significant. Goffman and Morris found that the pattern of journal usage in the Allen Memorial Library follows a Bradford distribution. Rao, through his analysis of circulation data collected from six Canadian University Libraries, proved that the rank distribution of transactions follow a Bradford distribution. Bradford's law is very much akin to the Pareto's law relating to wealth distribution and the 80:20 principle used in warehouse management and the Mandelbrot distribution.

Naturally this law is applied to study not only the scattering of publications, but also in other spheres of activity also. A study conducted by Garg and Lalitha Sharma of R & D indicators in Indian industry using Bradford's law bears testimony to this fact[31]. By analysing the R & D expenditure of 452 in --house R & D units in different sectors of the Indian industry, they had identified 19 in-house R & D houses as the core, 60 as the medium and the rest as small. As compared to medium and small-level in-house R & D units, there is a heavy concentration of manpower deployed, papers published, patents filed, processes/products developed in the core in-house R & D units. This shows the superiority of the core not only in the R & D expenditure but also in other yardsticks too.

2.2 *Zipf's Law*

Zipf's law can be effectively used in the generation of semi-automatic or automatic indexes useful for an information retrieval system. Its use has increased tremendously with the emergence of natural language indexing of textual matter especially in electronic form. Several studies aimed at finding out the pattern of frequency distribution of descriptors of a thesaurus and the distribution of indexing terms are available. A prominent one among them is that of Fedorowicz[32].

Zunde and Slamecka have developed a function for the optimum distribution of indexing terms by the number of postings[33]. This should make it possible to transmit information with maximum efficiency.

Zipf's law provides a measure of the richness in vocabulary of an author. This technique can be used for deciding the correct authorship of disputed works. For example, if there is difference of opinion as to the correct author of a work, the word predilections of the attributed authors can be analysed either manually or using a computer. Once the frequency of occurrence of favourite words are decided, the disputed text can be analysed to see similarity and thereby decide the author conclusively.

The law is also used for identifying words more frequently used in different foreign languages. These words are taught first in the instructional programmes of foreign languages.

Emile C White observes that the super-imposition of the Bradford distribution over the linear Zipf distribution, which demonstrates the emergence of more used and popular items may yield a technique to describe the pattern of book use by library patrons. She feels that applied to circulation data, these formulations can support such policies as shortened loan periods for heavily used books and the identification of a core collection[34].

2.3 Lotka's Law

Lotka's proposition led to a whole gamut of studies on scientific productivity. Such productivity studies have gained momentum in the post-second world war period. This in fact, has culminated in the rise of a new discipline called Scientometrics. Scientometrics is defined as the study of the measurement of scientific and technological progress. It provides an understanding of the structure of scientific activity, the disciplines being researched, the organizations involved, the strength and deficiency in the scientific groups and their communication channels at different levels of aggregation. It follows the trajectory of econometrics in the use of quantitative data, concepts and models and extensive use of mathematical and statistical techniques of modeling and data analysis.

Scientific productivity studies have been made from different angles. Impact of social change on scientific productivity, relationship of publication output on scientific recognition, identification of elites in different disciplines, occurrence of discoveries in different cultures etc. are some of the approaches made in this line.

An interesting study of scientific productivity made three decades back is that of Yuasa[35]. Yuasa's was a statistical study of the scientific achievements in various countries that showed the shifting of the world scientific dominance from one country to another. He found out that this dominance shifted from Italy to Britain, then to France, from France to Germany and finally to USA in the 20th Century.

Price, who had traced the development of science since Babylon and plotted the growth of big science from little science had observed that Lotka's law applied equally well to the productivity of scientists in the 17th as well as in the 20th century. This meant that majority of publications emanated from a handful of people. We have already seen this as Price's square root law. Similarly, the conclusion of an extensive review of early studies of scientific productivity made by Narin was that scientific talent was highly concentrated in a limited number of individuals[36].

For a comprehensive treatment of the application of bibliometric laws, readers are advised to refer to Egghe and Rousseau[37], albeit, abound in mathematical expressions.

2.4 Citation analysis

The primary function of citation is to provide "a connection between two documents, one which cites and the other which is cited"[38]. There are umpteen number of reasons for giving citations. Weinstock[39], Lipetz[40], Moravcsik and Murugesan[41], Hodges[42], Oppenheim and Renn[43], Finney[44], Frost[45] and Thorne[46] have all attempted to explore the possible reasons for giving citations. They include the positive and negative reasons for inclusion. However, it has to be conceded that if the reason is positive, there is bound to have some connection between the citing and cited paper. The first recorded citation analysis was a study by P.L.K. Gross and E.M. Gross published in 1927 in order to determine the journals to be subscribed to and the back volumes to be acquired for the Library of the Pomona College. They studied the citation frequency in the references given in the *Journal of the American Chemical Society*[47]. Citation analysis is very often fruitfully applied to derive the following benefits:

a. To lead the reader to further studies in the field

This is perhaps, the primary purpose of citations. Readers can verify the correctness of the information and thereby convince themselves.

b. For the preparation of bibliographies

The first use of citation indexing was made in *Shepherd's Citations* published in 1873. This technique of citation indexing has been perfected by Eugene Garfield and others since the early 1960s[48]. It is a fact that compilation of bibliographies in new fields is really difficult. In such circumstances, analysis of citations of articles may be the only way to gather information. The very fact that the citations have been verified, evaluated and recommended by authors who are experts in their own fields make them all the more acceptable for inclusion in a bibliography.

c. *To study the use pattern of different types of documents*

Citations given may be of books, journal articles, reports, standards, theses/dissertations etc. The relative use of each of these types can be ascertained based on the frequency of citations. For example, various citation studies have shown that journal articles are the most preferred source consulted by scientists since they constitute about 70-80% of the total citations. Similarly citation practices among social scientists indicate that they give equal importance to books and journals.

d. *To find out the relative use of different languages*

Since English has emerged as a world language, especially in science and technology, there is a predominance of English language publications in all branches. This can easily be understood from citation analysis. In the mid-sixties, for instance, the share of English language papers in Mathematics and Chemistry was more than 50 percent. Russian occupied the second position with about 20 percent followed by German and French.

Citation practices have also shown that the relative amount of literature in different subjects produced by different countries changes with time. It has been observed that German has declined very much in the 20th century, especially in the field of Chemistry where publications in this language reigned supreme.

e. *To study the use of literature from different countries*

From the citations, the country of their origin can be identified in all types of materials like journal articles, books, reports etc. In many subject areas, U.S. Publications are found to be used more heavily. In medicine, biochemistry, physiology and pharmacology, Sengupta had identified the leading role played by U.S. journals. Journals of U.K. occupied the second position, but they come nowhere near their American counterparts in the frequency of use. Similarly, Martyn and Gilchrist had found that in sixties one in every eight citations was to British publication[49]. Some of the user studies in India have shown that Indian publications are also equally cited in certain subjects.

f. To study the scattering of subjects

Studies about the dispersion or scattering of subjects in different sources as evidenced by citation analysis have brought out interesting results. For example,

i. Social science and arts subjects show a wider scatter of publications than the sciences.

ii. Research publications in technology show a greater dispersion than those in science.

iii. A new branch of science, especially an interdisciplinary one, shows a greater dispersion than an older branch of science.

iv. There can be differences in scatter between sub-fields within a subject as also among major subjects.

v. The rate of scatter within the same subject alters with time.

Meadows has summarized these findings[50].

g. To decide the obsolescence rate of documents in different subjects

Citations in subsequent literature and usage pattern in libraries are considered as two indicators of the obsolescence of literature. Analysis of citations by age of the cited document can show the useful life of a document. In order to measure the decay or obsolescence rate of documents, the concept of 'half life' has been borrowed from Nuclear Physics. Using this measure Burton and Kebler had suggested a range of half-lives for different subjects[51]. The fast growing subjects would have lesser half lives compared to established disciplines. The above study had shown the half-life of Metallurgical Engineering as 3.9 while that of Botany is 10 years. These time scales are highly useful in the planning of library holdings.

h. To determine the interdependence and lineage of subjects

The interdependence of basic and applied fields can be understood by citation studies. Establishment of this interdependence can be of use in the acquisition policy of special libraries or information centres. The analysis of citations of the *Annual Review of Medicine* for the years 1965-69 by I.N. Sengupta has established the contribution made by journals in the fields of biochemistry and physiology to the medical research[52]. Further studies by him have brought to light the mutual contribution of biochemistry, physiology and microbiology[53].

As far as lineage of subjects are concerned, Garfield's experiments in citation indexes have very much contributed in mapping the history of many of them.

i. To prepare ranked list of periodicals

Ranked list of periodicals can be prepared by two methods:

(1) by actual citation counting; and

(2) by counting the number of entries in indexing and abstracting periodicals.

In the first method, information is collected from the references cited in source articles. By studying the average number of citations, one can develop a list of cited journals in the ranked order.

In the second method, the number of items contributed by different periodicals during a specific period of time is calculated from the secondary source and the ranked list is prepared based on the productivity of journals. Such ranked lists are very often used as guidelines in the acquisition of periodicals and other materials in the library.

j. To study the rate of collaborative research

Collaborative research can be effectively measured from the number of authors in papers. Such studies can be conducted to understand global trends, national trends or trends in different subjects. Studies in this direction have indicated that collaboration varies from discipline to discipline, within the same discipline from time to time, and from country to country. However, the extent of collaboration may not be revealed from the citations. Efforts in this direction have been made by Ajiferuke et al who have attempted to define 'good collaboration' neasures[54]

k. For the analysis of scientific journals

Citation analysis provides a number of interesting and useful insights into the networking of journals. These insights are developed from five different citation measures, which are perfected by Institute for Scientific Information (ISI)[55].

i. Citation rate of a journal:

This is the number of times a journal has been cited. It can consist of all the references to the cited journal counting even duplicate references from the same source article as a seperate citation. It can also be calculated by counting only the number of source articles that cited the journal. A third method of calculating citation rate that is followed by ISI is by counting the number of references to the cited journal, but treating duplicate references from the same source article as only a single citation link.

ii. Impact factor:

Impact factor (IF) is the average citation rate of a journal's articles. It is basically a ratio between the citation rate of the journal and its citation potential. Citation rate is defined as the number of times cited, whereas citation potential is defined as the number of citable items published.

Therefore IF = the number of times a journal was cited/the number of citable items the journal published

Thus, the 1986 impact factor of Journal X would be calculated by dividing the number of all the Science Citation Index, Social Science Citation Index and Arts & Humanities Citation Index Source Journals' 1986 citations of articles journal X published in 1984 and 1985 by the number of items it published in 1984 and 1985.

Previously, it was not possible to calculate the IF of journals not covered by Science Citation Index. But now a new formula for the determination of IF for journals which are not incorporated in it has been worked out by B.K. Sen and others. The method is discussed in a short communication in *Journal of Documentation*, 45(2) 1989.

iii. Self-citing rate:

This is a measurement of the frequency with which journals cite themselves. It shows what percentage of a journal's references cite articles it published.

iv. Self-cited rate:

This, again, is a measurement of self-citation. It shows what percentage of citations received by a journal originated in articles published by the journal. These self-citation rates serve as indexes to the newness, size and isolation of the intellectual universe in which a journal operates.

v. *Immediacy index:*

This is a method of showing how rapidly the materials published by a journal are picked up and used. It is calculated by counting the number of citations received by articles in a journal during the year in which they were published.

The results of the citation measures carried out by ISI are published regularly in the Journal Citation Reports (JCR) of ISI.

Citation counts have also been stretched even to measure the productivity in other areas. As Broadus points out, "Over the last two decades, tabulation of citations have been used to measure the importance of academic departments, but especially of individual scholars and the contributions they have made to their respective fields" 56. In addition to the above areas of application, citation analysis has also led to the development of such concepts like bibliographic coupling put forward by Kessler[57] and co-citations by Small[58]. Co-citation is getting renewed attention now-a-days.

3. LIMITATIONS IN APPLICATION

3.1 Bibliometric Laws

Though most of the studies tend to support the Bradford distribution, some other researchers could not get satisfactory results. Groos found that the scatter of research papers among Physics journals deviated from that predicted by Bradford's law[59]. Chonez tested a large number of different areas and found that the law applied in a very small proportion of them[60]. Out of fifty bibliographies studied by him, only six followed the law. Therefore, he calls the law pseudo-scientific. However, much of the deviation is attributed to the problems of initial data collection.

In the case of Lotka's law, it was found to fit in most cases. However, the value of the index 'n' was found to vary for different groups of scientists. For example Dobrov and Korennoi found that Lotka's value can be used in measuring productivity in botany, but not in electronics[61]. Another problem with Lotka's law is that it totally ignores the potential authors who have not produced any publication so far. Because of these limitations, the empiric nature of these laws are questioned.

3.2 Citation studies

We have seen that the very logic of depending on citations which are given according to the whims and fancies of authors has been questioned by many researchers. They believe that any result that is obtained through such a study of citation leads to erroneous conclusions. Maurice B Line is a strong votary of this point of view.

The common arguments levelled against citation analysis are the following:

i. Negative citation ie. citing a paper just to repudiate it. The result is that controversial papers will get more citations than really worthwhile papers.

ii. Too much of self-citation and in-house citation.

iii. Practice or citing only to get the favours of the powerful or to appease others.

iv. Citations given just to dress up the paper.

v. Variation of citation rate during life time of paper.

vi. Variation of citation rate with type of paper and speciality.

It is a fact that there are extraneous considerations in giving citations. But that does not totally undermine the value of them. As Sengupta argues, "the number of scientists who are capable of doing such malpractices are not significantly high to make citation studies unworkable or misleading"[62]. Further, it is too much to question the honesty and integrity of the majority of scientists.

4. CONCLUSION

Bibliometrics has emerged as the most active field of Library and Information Science. It is estimated that literature on this topic occupies more than 25 percent of the total contribution in L & IS. Citation analysis studies form a major portion of it. The rate of growth of the literature is obvious from the number of entries of the bibliographies on the subject. Pritchard and Wittig could locate 600 entries on the subject while covering the period 1874 to 1959[63]. When another bibliography was compiled by Hjerppe in 1980, the number of entries rose to 2032[64]. He could add 518 items more within two years. Peritz, who analysed the period from 1960 to 1985 could count 3225 items[65]. To this are added hundreds of studies every year published in journals exclusively devoted to the field as well as the other L & IS journals and the proceedings of many conferences including the biennial International Conferences on Bibliometrics and Informetrics (so far held in 1987, 1989, 1991, 1993 & 1995).

A major portion of these studies pertains to the application of bibliometric laws and models. However, there is a long way to go in achieving perfection in the studies. Even the widespread use of computers for retrieval, counting and analysis are unlikely to achieve perfection in the studies. The changes that are frequently occurring in the publication practices are likely to complicate the studies in future. In such circumstances, it is advisable to consider the results of such studies as mere guidelines rather than ends in themselves.

REFERENCES:

1. Cole, F.J. and Eales, N.B. : The history of comparative anatomy. Part I: a statistical analysis of literature. Science Progress 11, 1917: pp. 578-596.
2. Hulme, E.W. : Statistical bibliography in relation to the growth of modern civilization. London: Grafton, 1923.
3. Lotka, A.J. : The frequency distribution of scientific productivity. Journal of the Washington Academy of Sciences 16, 1926: pp. 317-323.
4. Zipf, G.K. : Human behaviour and the principle of least effort. Cambridge, Massachusetts: Addison-Wesley, 1949.
5. Bradford, S.C. : Documentation. London: Crossby Lockwood, 1948.
6. Ranganathan, S.R. : Librametry and its scope. DRTC Seminar 7. Paper DA. Bangalore: DRTC, 1969: pp. 285-301.
7. Price, Derek J de Solla : Networks of scientific papers. Science 149, 1965: pp. 510-515.
8. Kessler, M.M. : Bibliographic coupling between scientific papers. American Documentation 14(1) 1963: pp. 10-25.
9. Garfield, Eugene : Citation analysis as a tool in journal evaluation. Science 178(4060) Nov. 1972: pp. 471-479.
10. Goffman, W and Warren, K.S. : Scientific information systems and the principle of selectivity. New York: Praeger Press, 1980.
11. Pritchard, Alan : Statistical bibliography or bibliometrics. Journal of Documentation 25, 1969: pp. 348-349.

12. Egghe, L. : Methodological aspects of bibliometrics. Library Science with a slant to Documentation and Information Studies. 25, 1988: pp. 179-191.

13. Narin, F and Moll, Joy K. : Bibliometrics: In Annual Review of Information Science and Technology Vol 12. New York: Knowledge Industry Publications, 1977.

14. Price, Derek J de Solla : A general theory of bibliometrics and other cumulative advantage process. Journal of American Society for Information Science 27(5) 1976: pp. 292-306.

15. Meadows, A.J. : Communication in Science. London: Butterworths, 1974.

16. Fairthorne, R.A. : Empirical hyperbolic distributions (Bradford-Zipf-Mandelbrot) for bibliometric description and prediction. Journal of Documentation 25(4) 1969: pp. 319-43.

17. Bookstein, Abraham : Explanations of the bibliometric laws. Collection Management 3, 1979: pp. 151-162.

18. Chen, Y.S. and Leimkuhler, F.F. : A relationship between Lotka's law, Bradfrod's law and Zipf's law. Journal of the American Society for Information Science 37, 1986: pp. 307-314.

19. Price, Derek J de Solla : Some remarks on elitism in information and the invisible college phenomenon in Science. Journal of American Society for Information Science 22, 1971: pp. 74-75.

20. Egghe, L and Rousseau, R. : Introduction to informetrics. Amsterdam: Elsevier, 1990. p. 377.

21. Garfield, Eugene : The mystery of transposed journal lists. Current Contents (Life Sciences) 14(31) 1971: pp. 5-6.

22. Sengupta, I.N. : Growth of biochemical literature. Nature 244, 1973: pp. 75-76.

23. Ravichandra Rao, I.K. : In Workshop on Scientific Communication and Biliometrics. Calcutta, 1988.

24. Brookes, B.C. : Numeric methods of bibliographic analysis. Library Trends 22(1) July 1973: pp. 18-43.

25. White, Emilie C. : Bibliometrics: from curiosity to convention. Special Libraries Winter 1985: pp. 35-42.

26. Ferrante, Barbara Kopelock : Bibliometrics: access in the library literature. Collection Management 2(3) 1978: pp. 199-204.

27. Burton, Hilary D. : Use of a virtual information system for bibliometric analysis. Information Processing and Management 24(1) 1988: pp. 39-44.

28. Lancaster, F.W. : If you want to evaluate your library... 2nd ed. London: Library Association, 1993. pp. 285-290.

29. Goffman, William and Morris, Thomas G. : Bradford's law and library

acquisitions. Nature 226, 1970: pp. 922-923.

30. Ravichandra Rao, I.K. : Document and user distribution in transaction records of Canadian University libraries. Doctoral Thesis, Ontario, Canada: University of Western Ontario, 1981. (Quoted in Ravichandra Rao, I.K. Quantitative methods in Library and Information Science. New Delhi: Wiley Eastern, 1983).

31. Garg, K.C. and Lalitha Sharma : A study of R&D indicators in Indian industry using Bradford's law. Journal of Scientific and Industrial Research 53, Feb. 1994: pp. 81-87.

32. Fedorowicz, J. : The theoretical foundation of Zipf's law and its application to the bibliographic database environment. Journal of American Society for Information Science 33, 1982: pp. 285-293.

33. Zunde, Pranas and Slamecka, Vladimir : Distribution of indexing terms for maximum efficiency of information transmission. American Documentation 18(2) 1967: pp. 104-108.

34. White, Emilie C. : Bibliometrics: from curiosity to convention. Special Libraries, Winter 1985: pp. 35-42.

35. Yuasa, M. : Centre of scientific activity, its shift from the 16th to 20th century. Japanese Studies in the History of Science, No. 1, 1962: pp. 527.

36. Narin, F. : Evaluative bibliometrics: the use of publication and citation analysis in the evaluation of scientific activity. New Jersey: Cherry Hill, 1976.

37. Egghe, L and Roussean, R. : Introduction to informetrics. Amsterdam: Elsevier, 1990. pp. 361-372.

38. Martyn, J. : Citation analysis. Journal of Documentation 31(4) 1975: pp. 290-297.

39. Weinstock, M. : Citation indexes. In Encyclopedia of Library and Information Science. Vol. 5. New York: Marcel Dekker, 1971.

40. Lipetz, B.A. : Improvement of the selectivity of citation indexes to science literature through the inclusion of citation relationship indicators. American Documentation 16(2) 1965: pp. 81-90.

41. Moravcsik, M.J. and Murugesan, P. : Some results on the function and quality of citations. Social Studies of Science 5, 1975: pp. 86-92.

42. Hodges, T. : Citation indexing: its potential for bibliographic control. Ph.D. Thesis. Barkeley: University of California, 1978.

43. Oppenheim, C and Renn, S.P. : Highly cited old papers and the reasons why they continue to be cited. Journal of American Society for Information Science 29(5) 1978: pp. 225-231.

44. Finney, B. : The reference characteristics of scientific texts. M.Sc. Dissertation. London : Centre for Information Science, City University, 1979.

45. Frost, C. : The use of citations in literary research: Preliminary classification

of citation functions. Library Quarterly 49(4) 1979: pp. 399-414.

46. Thorne, F.C. : The citation index: another cause of spurious validity. Journal of Clinical Psychology 33, 1971: pp. 1157-1161.

47. Groos, P.L.K. and Groose, E.M. : College libraries and chemical education. Science 66, 1927: pp. 1229-1234.

48. Garfield, Eugene and Sher, I.H. : New factors in the evaluation of scientific literature through citation indexing. American Documentation 14(4) 1963: pp. 195-201.

49. Martyn, J and Gilchrist, A. : An evaluation of British scientific periodicals. London: Aslib, 1968.

50. Meadows, A.J. : Communication in Science. London: Butterworths, 1974.

51. Burton, R.E. and Kebler, R.W. : The half-life of some scientific and technical literatures. American Documentation 11(1) 1960: pp. 18-22.

52. Sengupta, I.N. : Impact of scientific serials in the advancement of medical knowledge: an objective method of analysis. International Library Review 4, 1972: pp. 169-195.

53. Sengupta, I.N. : Bibliometric research: growth of biomedical literature. Vol. 1, Calcutta: SBA Publications, 1988.

54. Ajiferuke, I et al. : Collaborative coeffieient: a single measure of the degree of collaboration in research. Scientometrics 14, 1988: pp. 421-433.

55. Garfield, Eugene : Citation indexing - its theory and application in science, technology and humanities. New York: John Wiley, 1979.

56. Broadus, Robert N. : The application of citation analysis to collection building. In Advances in librarianship. Vol 7. New York: Academic Press, 1977.

57. Kessler, M.M. : Bibliographic coupling between scientific papers. American Documentation 14(1) 1963: pp. 10-25.

58. Small, H. : Co-citation in the scientific literature: a new measure of the inter-relationship between two documents. Journal of the American Society for Information Science 24, 1973: pp. 266-269.

59. Groos, O.V. : Bradford's law and the Kennan-Atherton data. American Documentation 18, 1967: p. 46.

60. Chonez, A. : La dispersion de la literature periodique en science de l' information. Documentaliste 11(4) 1974: pp. 175-184.

61. Quoted in Meadows, A.J. : Communication in science. London: Butterworths, 1974. p. 193.

62. Sengupta, I.N. : Bibliometric research: growth of biomedical literature. Vol 1. Calcutta: SBA Publications, 1988.

63. Pritchard, A and Wittig, G. : Bibliometrics: a bibliography and index (1874-1959). Vol. 1. Wartford, U.K: AAIM Books, 1960

64. Hjerppe, R. : A bibliography of bibliometrics and citation indexing and analysis. Stockholm, Sweden: Royal Institute of Technology Library, 1980.

65. Peritz, B.C. : A classification of citation roles for the social sciences and related fields. Scientometrics 5(5) 1983: pp. 303-312.

4

FINANCIAL PROBLEM OF PERIODICAL ACQUISITION:

A MATHEMATICAL SOLUTION

DR. MANORAMA SRINATH

0. INTRODUCTION

Periodicals are purvayers of information which are essential resources for carrying out research and teaching. Faculty and students of Library & Information Science, depend heavily on these resources, which are becoming costlier day by day. Escalation of cost of periodicals is a persisting problem of every librarian. Every year the librarians are forced to cut down certain periodicals due to the ever increasing cost of periodicals and as a result there is no consistency in periodical subscription.

This article gives an account of the budget sanctioned for the Department of Library and Information Science, University of Madras, and the details regarding the increase of cost of periodicals subscribed for the same department for a period of six years. Further, this article stresses the need to adopt a definite policy for the budget allocation. For this a mathematical equation has been worked out in this article which determines the average increase of the cost for a given number of periodicals.

1. PROBLEM OF PERIODICAL ACQUISITION

The general procedure in University libraries is to allocate certain amount for both periodicals and books of which more than 70% of amount will be spent for periodical acquisition. Even after spending more than 70%, the librarians often incur the displeasure of the users by discounting the subscription for some of the costly periodicals. Though the authorities of the University libraries are aware of these facts they are unable to rectify this situation due to two factors. The first factor is that the allocation of funds for the departments depends on the economic conditions of the University which in turn influenced by the revenue got by the regular income and the grants. The revenue got by the university may fluctuate due to various reasons and this fluctuation automatically reflects of the allocation of library funds.

Another significant factor which has been simply ignored is that the funds are allocated only on the basis of whole economic condition and not on the basis of increasing cost of periodicals. In other words the process of allocation of funds has neither been well planned nor a definite policy has been adopted.

(Table-1)

Cost of Periodicals from 1990-1995

	1990	1991	1992	1993	1994	1995	Total
1. Data Base Magazine	- -	- -	- -	- -	448	4234.50	8682.40
2. IFLA	- -	- -	- -	- -	3577.80	3631.20	7209
3. Information Library Manager (Logistics Inf. Man)	4478	D	D	24861.41	D	D	29339.41
4. Information Services & Use	2624	3702	5104	7449	D	D	18879
5. Infor. Technology & Libraries	- -	- -	1375	1749	1992	1896	7012
6. Inter. Cataloging & Bib. Control	- -	- -	- -	- -	1580.80	1799	3379.80
7. Inter. Information and Libraries	- -	- -	- -	6360	- -	- -	6360
8. Inter. Information Comm. & Education	180	225	280	- -	328	340	1353
9. Inter. Library Review	2850	3200	5000.25	- -	- -	- -	11050.25
10. J. of Information Science	2120	2996	4878.60	7086.10	- -	- -	17080.70
11. J. of Librarianship (J of Librarianship & Infor, Science)	1534	1947	2766	3367	- -	- -	9614
12. J. of Library & Infor. Science	80	80	80	80	- -	- -	320
13. Lib. Hi Tech. J.	- -	- -	- -	- -	2954	2812.40	5766.40
14. Library Resources & Tech. Services	- -	945	1512	1749	2158	2054	8418
15. Library Trends	922.20	1228	1787	2130	- -	- -	6067.20
Total	14788.20	14323	22782.85	54831.51	17038.60	16767	

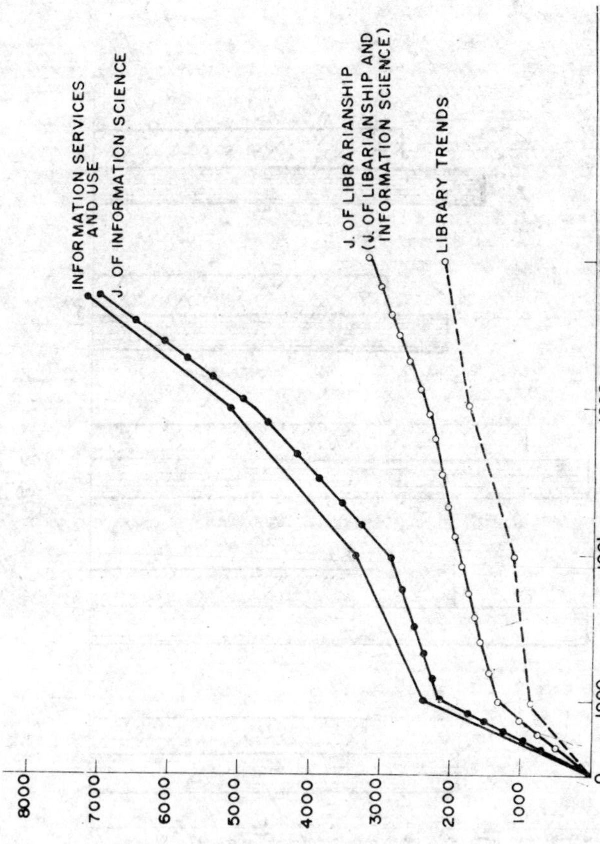

8000
7000
6000
5000
4000
3000
2000
1000
0
1990
1991
1992
1993
INFORMATION SERVICES AND USE
J. OF INFORMATION SCIENCE
J. OF LIBRARIANSHIP (J. OF LIBARIANSHIP AND INFORMATION SCIENCE)
LIBRARY TRENDS

Fig. 1

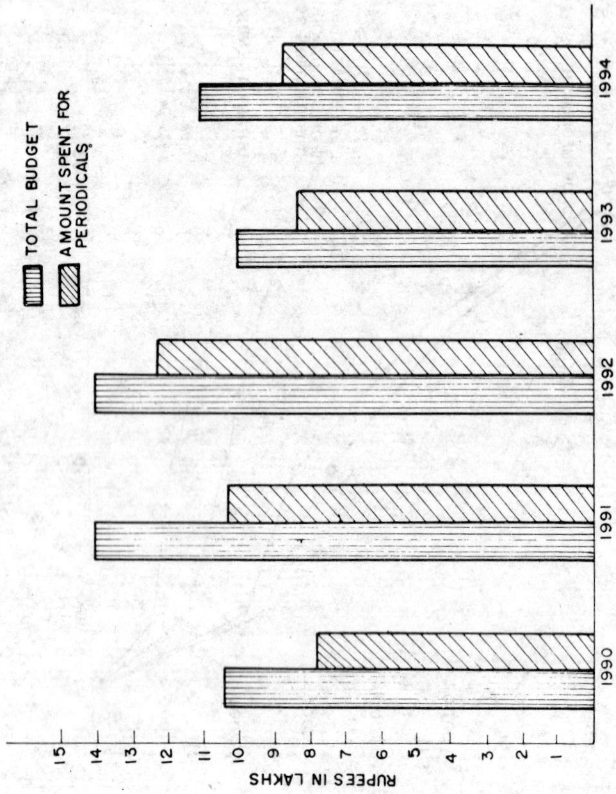
TOTAL BUDGET
AMOUNT SPENT FOR PERIODICALS
RUPEES IN LAKHS
15
14
13
12
11
10
9
8
7
6
5
4
3
2
1
1990
1991
1992
1993
1994

Fig. 2

2. DATA ANALYSIS

Table 1 gives a clear picture of the cost of periodicals from 1990 to 1995. The table shows that there is no consistency in periodicals acquisition. There are two reasons for this. One reason is that either the periodicals are not received even after paying the subscription or the subscription would have been stopped due to the increase of the cost of periodicals and lack of funds. The table shows that for nine periodicals Rs. 22,782.85 has been spent in 1992 & in 1993, Rs. 54,831.51 has been spent for same number of periodicals.

According to the data given in the graph (Graph) it is evident that the cost of two journals have increased considerably. For example the cost of Information Services and use has increased from Rs. 3702 to Rs. 7449, in three years. The difference of cost is Rs. 3747. Similarly the cost of Jl. of Information science has increased from Rs. 2996 to Rs. 7086. The difference of cost is Rs. 4090.

The average cost of four periodicals for three years has been calculated as Rs. 2468.25(1991), Rs. 3633.90 (1992), Rs. 5.008 (1993). The percentage increase of cost for four periodicals for 1993 has been calculated as 147.22 and for 1993 it was 137.81. In general, the percentage increase of cost of periodicals every year is more than 100%.

The bar diagram indicates that there is no relevance at all, between the budget allocation and the increase of cost of periodicals.

As per the bar diagram the budget amount sanctioned for the Department of Library and Information Science remained the same for a period of 3 years; i.e. from 1991 to 93 only Rs. 2000 was sanctioned. but the actual amount spent for subscribing nine periodicals for 1992-93 has been indicated as Rs. 54,831.51. Eventually the extra amount was spent from the funds allocated for the Main library. In 1992 & 93 nine periodicals were subscribed but the difference of cost is more than thirty two thousands.

In spite of the sharp increase in the cost of periodicals the budget amount remained the same and only in 1994 & 95 two thousand rupees have been increased in the budget allocation. As a result of this, the subscription for costly periodicals has been discontinued and orders were placed only for selected periodicals.

Hence the study indicates the following factors:-

1. The percentage increase of cost of periodicals every year is more than 100%.

2. The budget allocation has not been increased proportionately to the rise of cost of periodicals.

3. So there is no consistency in subscribing the periodicals literature.

3. FISCAL ALLOCATION: A MATHEMATICAL SOLUTION

To ensure a steady supply of periodicals it is essential to adopt a definite policy in allocating funds for the resources. The policy should stress that the increase in the budget amount should be based on certain ratio, which takes care of the average increase of cost of the given number of periodicals.

To determine the ratio a mathematical equation is given below:

C = Cost of periodicals

AC = Average cost of periodicals

PC = Percentage increase of cost of periodicals

Percentage to be added in the current year budget

$$= \frac{PC1 + PC2 = PC3}{2}$$

Extra amount or rupees to be added in the next year budget =

$$Rs. = \frac{C3 \times PC3}{100}$$

(Using the above equation the budget amount for 1994 has been calculated for four periodicals*, which are subscribed to the department of Library and Information Science, University of Madras. The budget to be allotted for 1996 could not be calculated because the data regarding the cost of periodicals for 1995 received at the time of writing this article was not quite complete because the supplementary bills were not yet received by the library.)

* 1. Information services and use

* 1. Information services and use

* 2. Journal of Information Science

* 3. Journal of Librarianship and Information

* 4. Library trends

1. The total cost of four periodicals for three years:

   For 1991 = C1 = Rs. 9,873

   For 1992 = C2 = Rs. 14,535.60

   For 1993 = C3 = Rs. 20,032.10

2. The average cost of four periodicals for three years:

   For 1991 = AC1 = Rs. 2468.25

   For 1992 = AC2 = Rs. 3633.90

   For 1993 = AC3 = Rs. 5008

3. Percentage increase of cost of four periodicals

For 1992 PC1 $= \frac{100 \text{ X } AC2}{AC1}$

$= \frac{100 \text{ X } 3633.90}{2468.25}$

$= 147.22$

For 1993 PC2 $= \frac{100 \text{ X } AC3}{AC2}$

$= \frac{100 \text{ X } 5008}{3633.90}$

$= 137.81$

4. Percentage for 1994 PC3

$$= \frac{147.22 + 137.81}{2}$$

$= 142.51$

5. Extra amount which should have been added for 1994

$$= \frac{20032.10 \times 142.51}{100}$$

$$= 28547.74$$

As per the above calculation for four periodicals Rs. 28547.74 should have been added with the budget amount of the previous year.

4. CONCLUTION

If the above formula is adopted while sanctioning the budget for periodical acquisition the staff and students of Library Science department will be assured of consistence supply of periodical information. Because of the increase of cost of periodicals and the limited budget the librarian is forced to spend the budget allotted for books to meet the supplementary bills of periodicals and the department is deprived of buying required new books. Hence this article tries to draw the attention of the concerned authorities and emphasizes the need to adopt this formula while sanctioning the budget for the department of Library and Information Science.

REFERENCES:

1. Egghe, L : Methodological aspects of bibliometrics. Library Science with a slaut to Doc. 25. pp. 179-191.
2. Garfield, E. : Citation analysis as a tool in journal evaluation Science 178 (4060) Nov. 1972. pp. 471-479.
3. Meadows, AJ : Communication in Science London, Butterworths, 1974.

5

SOCIAL SCIENCE BOOKS IN MALAYALAM: A BIBLIOMETRIC PERSPECTIVE

BEENA-S

0. INTRODUCTION

The literature of the social science with its documentary interconnections represent only part of the total communication system of the social scientists. Even though a number of newer media or non-book materials, have emerged as information carriers, majority of the social scientists still rely on the traditional literary forms of communication, inspite of the weakness of the system.

Books are the end products of the intellectual process of disciplined minds of men and women. Book is a repository of information, knowledge and media of communication from mind to mind. Without books the growth of knowledge and its extension would have been far behind or even stagnant. All educational activities-whether formal or informal-have to depend on the printed books. Right from the primary level to the higher education, both students and teachers choose and read the books. Even research work cannot completely be independent of books. Social sciences have a very wide connotation and appropriately a very large number of books, including several of the valuable classical titles are in use even today. Many new books are emerging in this field which are very valuable to people especially to students and researchers.

Even though there are a lot of social science books in the English language their number is comparatively low in Malayalam. This is mainly because of the widespread use of English as the medium for study and teaching, especially at the higher levels of education. But today the shift in emphasis to Malayalam could be evinced. The government's initiative to switch over to Malayalam as the official language gave a fillip to the production of more books in the language. Therefore, it is worthwhile to study the rate of production of social science books and other related aspects of book production in Malayalam.

Inspite of the importance attached to books in social sciences, very few studies are available about their production and growth. This is more so in the case of Malayalam. For effective communication of information, the language of communication should be in the language most familiar to the users.

1. SCOPE OF SOCIAL SCIENCES

The subjects that cannot be easily incorporated under either Sciences or Humanities are included in Social Sciences.

For the purpose of the study the subjects that are taken to the Social Sciences are Education, Geography, History, Political Science, Economics, Sociology and Law. In order to arrive at a conclusion as to the disciplines that can be brought under the purview of Social Sciences, an extensive survey of definitions was made. Since the study was about books in a regional language, it was found that the pattern followed by Colon classification is most suited which cover these seven subjects under the purview of Social Sciences.

2. OBJECTIVES

The major objectives of the study are:

(a) to assess the share of social science books in Malayalam;

(b) to find out the social science disciplines in which there is need for publication of more books;

(c) to analyse the titles that were published during 1981-93 using various bibliometric variables;

(d) to assess the share of translations from other languages in various disciplines;

(e) to ascertain the rate of increase in cost over the period of 13 years;

(f) to locate the major publishers of social science books in Malayalam and the centres of production.

3. METHODOLOGY

Methodology applied in the present study involves literature survey and bibliometric analysis

3.1 Literature survey

Literature on social sciences including books and periodicals were consulted. In addition, the C.D-Rom data bases, LISA PLUS and ERIC were also searched for relevant studies. But a bibliometric study of social science books as envisaged here could not be located.

For the collection of data various sources have been consulted. The most important among them are:

(1) Malayala Grantha Soochi, Edited by K.M. Govi

(2) Indian National Bibliography (Malayalam) Cumulative volume for ten years (1980-90) published by the Central Reference Library, Calcutta.

(3) The catalogue of Malayalam Books in the British Museum edited by Albertine Gaur

(4) The catalogues of major publishers in Malayalam

(5) The catalogues of important libraries like Kerala University Library, the State Central Library, the Kerala Sahitya Academy Library etc.

(6) Accession Register of the Kerala Sahitya Academy Library, Trissur.

The scope of the study is confined to a limited span of thirteen years ie. 1981-1993. Before 1981 more than three thousand Malayalam books were published in social sciences.

3.2 Bibliometric Analysis

For the detailed analysis, information about the social science books published in Malayalam from 1981 onwards upto the end of 1993 were collected. It was found that there were altogether 1390 titles. These include original works, translations, edited works and compilations. Editions and reprints are also included. These titles were analysed using several variables used for bibliometric analysis.

4. ANALYSIS OF DATA

4.1 Subject-wise distribution of Social Science books

In order to find out the number of books in each subject coming under the purview of Social Sciences, a subject-wise analysis was made.

As stated earlier seven subjects were considered under the generic term social sciences. The subjectwise break-up of books is given in the following table.

Table-1 Subject-wise distribution of Social Science books in Malayalam from 1981-1993

Subjects	Total books	Percentage
Education	74	5.32
Geography	173	12.45
History	402	28.92
Political Science	286	20.57
Economics	142	10.22
Sociology	289	20.79
Law	24	1.73
Total	1390	100.00

The table shows that history books come first constituting about 29 percent of the total social science books. Sociology books occupy the next position with a percentage of 20.79. Even though books on Political Science come only after Sociology

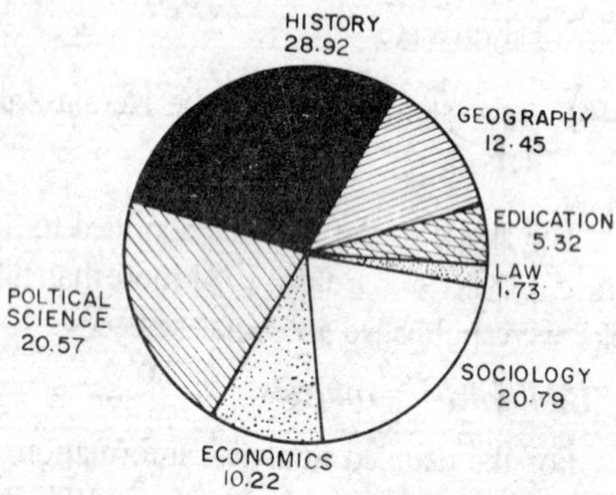

there is not much difference in their number. These three subjects constitute more than 70 percent of the total. Law books and books on Education are comparatively less in Malayalam language.

4.1.1 Subdivisions of Subjects

With a view to identify the major areas in which books have come in Malayalam in each subject, a detailed analysis was made. It was found that general books on each subject constituted the major portion in almost every social science subject except Geography and History. Books on travel constitute the major portion in Geography while biographical sketches form the highest number in History. In Political Science books on Marxism, Communism and Socialism together constitute nearly one-third. Caste, social reforms and women's studies are the subject areas in which Sociology books are mostly found. Surprisingly, there are a number of books on topics connected with women which could be included under the category 'Women's Studies'. It is an indication of the awareness among the people about the need for giving due attention to women.

4.1.2 Biographies

General biographies are not included in the study. Only those biographies that come under the purview of social sciences are considered. These constitute 14.24 percent of the total Malayalam books in the social sciences. A subject-wise distribution of the biographies is given below.

Table-2 Biographies related to Social Science Subjects

Subjects	Total Books	Biographies	Percentage of each subject
Education	74	- -	- -
Geography	173	- -	- -
History	402	134	33.33
Political Science	286	29	10.14
Economics	142	2	1.64
Sociology	289	33	11.42
Law	24	- -	- -
Total	1390	198	14.24

From the table it is clear that biographies are more in History than in any other subject. The life history of important personalities like rulers and national leaders constitute the biographies in History. In political Science most of them are that of revolutionaries and leaders of political parties. The life history of social reformers and leaders in the society form biographies in Sociology.

4.1.3. Multi-volumed Books

Only 1.73% of books are multi-volumed. Table 3 shows the number of multi-volumed books in each subject.

Table-3 Multi-volumed books in each subject

Subjects	Total Books	Multi-volumed Books
Education	74	- -
Geography	173	2
History	402	11
Political Science	286	8
Economics	142	1
Sociology	289	4
Law	24	- -
Total	1390	24

Multi-volumed books are totally absent in Education and Law. They are more in History and Political Science.

4.1.4 Year-wise production of Social Science Books

It is very important to find out whether the publication of social science books is increasing or decreasing year by year. The results of the analysis is carried out in the following table. (Table 4)

It can be noted from the table that the highest number of books is produced in the year 1990 (10.79%). In that year the production rate in all subjects except Education was very high. Next comes the year 1987 (9.27) and 1986 (9.06%) respectively. The same percentage of books are produced in the years 1981, 1984 and 1985 (8.49%). But after 1990 a decreasing tendency in book production can be noted. It is the lowest in the year 1993. This may be due to the fact that all the books published in 1993 might not have reached the market places and libraries in time.

With regard to subject-wise publication of books, the highest number in Education was published in the year 1982 (16 books). 1982 stands next (9). The number of contribution went down in the following years. In 1991 there was a slight increase. It increased upto 7. But in the next two years it again decreased. In 1993 it came down to 2. It can be observed that the growth of Malayalam books in Education was very unstable.

In Geography the highest number of books were published in the year 1985 and 1990. The production of books upto 1990 is above the average of 13 except in 1982, 1983 and 1989. But after 1990 there is a sharp decrease. This may be due to the fact that in geography

Table-4 Year-wise production of Social Science Books in Malayalam

Subjects	1981	1982	1983	1984	1985	1986	1987	1988	1989	1990	1991	1992	1993	Total	Average Books per Year
Education	9	16	6	4	4	6	3	6	5	3	7	3	2	74	5.7
Geography	18	8	12	16	22	15	14	15	9	22	11	6	5	173	13.3
History	25	31	25	32	38	37	37	46	37	52	13	20	9	402	30.9
Political Science	24	18	19	23	13	25	32	27	30	29	16	23	7	286	22.0

Subjects	1981	1982	1983	1984	1985	1986	1987	1988	1989	1990	1991	1992	1993	Total	Average Books per Year
Economics	16	14	5	16	16	15	15	7	9	12	3	9	5	142	10.9
Sociology	24	17	13	25	28	25	19	21	28	37	18	9	289	22.2	
Law	2	1	1	2	-	-	3	3	4	4	3	1	-	24	1.8
Total	118	105	81	118	118	126	129	123	115	150	90	80	37	1390	106.8
Percentage	8.49	7.55	5.83	8.49	8.49	9.06	9.28	8.85	8.27	10.79	6.48	5.76	2.66	100	

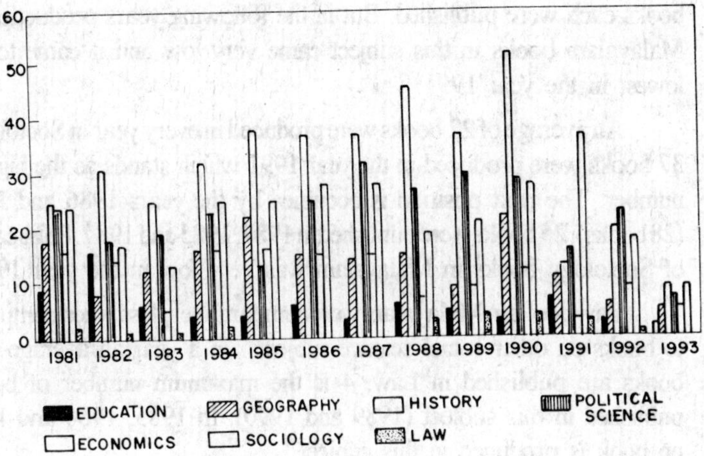

Fig. 2

the majority of the books are travelogues and after 1990 the writing of travelogues declined considerably.

The average books produced in a year is comparatively more in history than in any other subjects. In history only in 1991 and 1993 did the book production came very low. Book production was at its peak in 1990 (52 books). Here also after 1990 there was decline. A rise in publication can be detected from 1984 onwards upto 1990.

The highest number of Malayalam books in Political Science were published in the years 1987 (32) and in 1989 (30). In 1990, 29 books were published. Publication is low in the years 1985, 1991 and 1993.

16 was the highest number of books produced in Economics. The years 1981, 1984 and 1985 share this number. In 1986 and 1987, 15 books each were published. But in the following years production of Malayalam books in this subject came very low and it came to the lowest in the year 1991.

An average of 22 books were produced in every year in Sociology. 37 books were produced in the year 1991 which stands as the highest number. The next position is occupied by the years 1986 and 1990 (28 books). 25 books were published in 1984, 1985 and 1987. Publication of Sociology books in Malayalam was very low in the year 1993.

Law books in Malayalam were comparatively less when compared to books on other social science subjects. In a year an average of 2 books are published in Law. 4 is the maximum number of books published in this subject (1989 and 1990). In 1985, 1986 and 1990 no book is produced in this subject.

4.1.5 Authorship Pattern

Various bibliometric studies in Science and Technology have shown that there is considerable change in the authorship pattern of books and journal articles. The finding is that two author, three author and multiple author papers and books are on the increase. This was found to be true in Physics, Chemistry, Botany, Zoology etc. In order to find out whether this is true of social sciences also, the authorship pattern of books was analysed.

4.1.5.1 Type of authorship

For the purpose of analysis, three categories of authors were considered - single author, two author and multiple author. Details of the compiled works, edited works and works of corporate authors were also collected. The results are presented in Table 5.

The above table shows that 88.56 percent of the books are written by single authors. Books by one author are found to be highest in geography. Books written by more than one author are totally absent in Law. Books by more than two authors form only 1.44 percent of the total. Joint authorship is more in Political Science. Joint authorship is more in Political Science. Compiled and edited works together constitute only 1.87 percent. In other words, compilations are few and

Table-5 Authorship pattern in different subjects

Nature of Authorship	Education	Geography	History	Political Science	Economics	Sociology	Law	Total
Single Author (79.73*)	59 (95.95*)	166 (90.55*)	364 (82.52*)	236 (88.73*)	126 (90*)	260 (83.33*)	20 (88.56*)	1231
Two Authors (4.04*)	3 (1.16*)	2 (2.49*)	10 (3.50*)	10 (2.82*)	4 (2.42*)	7	- (2.59*)	36
Multiple Authorship	1 (1.35*)	1 (0.58*)	2 (0.50*)	7 (2.45*)	5 (3.52*)	4 (1.38*)	-	20 (1.44*)
Compilers (1.35*)	1	(0.25*)	1 (1.40*)	4	- (0.35*)	1	- (0.51*)	7
Editors (6.76*)	5 (0.58*)	1 (1.49*)	6 (0.67*)	2 -	- (1.73*)	5	- (1.36*)	19
Corporate Bodies	1 (1.35*)	-	2 (0.50*)	7 (2.45*)	-	3 (1.04*)	-	13 (0.94*)
Books without Proper author	4 (5.41*)	3 (1.73*)	17 (4.23*)	20 (7*)	7 (3.11*)	9 (16.67*)	4 (4.6*)	64
Total (100*)	74 (100*)	173 (100*)	402 (100*)	286 (100*)	142 (100*)	289 (100*)	24 (100*)	1390

Table-6 Distribution of authors and books in different subjects

No. of	Number of Authors							
Books	Eduction	Geography	History	Political	Economics Science	Sociology	Law	Total
1	44	103	217	171	113	212	16	876
2	6	19	34	28	11	19	2	119
3	1	6	17	3	3	10	-	40
4	1	1	1	2	1	-	-	6
5	-	-	3	1	-	1	-	5
6	-	1	1	-	-	1	-	3
7	-	-	1	-	1	-	-	2
8	-	-	1	-	-	-	-	1
More than 8	-	-	-	1	-	-	-	1

Table-7 Ten Major Publishers of Social Science books in Malayalam and the number of books

Sl. No.	Name of the Publisher	Number of books in each Social Science Subject							Total	Rank
		Education	Geography	History	Political Science	Economics	Sociology	Law		
1.	Chintha Publishers, Thiruvananthapuram	4	9	27	62	20	22	-	144	1
2.	Bhasha Institute Thiruvananthapuram	11	1	22	12	41	15	3	105	2
3.	D.C. Books Kottayam	-	25	36	17	5	17	1	101	3
4.	Sahitya Pravarthaka Sahakarana Sanghom (SPCS), Kottayam (including National Book Stall)	6	27	23	4	2	11	3	76	4
5.	Prabhatham, TVM	1	3	16	31	4	10	2	67	5

Sl. No.	Name of the Publisher	Number of books in each Social Science Subject								
		Education	Geography	History	Political	Economics Science	Sociology	Law	Total	Rank
6.	Indian Atheist Publishers, New Delhi	-	4	7	11	1	15	1	39	6
7.	Vidyarthimithram, Kottayam	12	3	13	1	2	4	-	35	7
8.	Kerala Sasthra Sahitya Parishath, TVM, Kollam, Trissur, Kozhikeds	4	2	3	6	10	9	1	35	7
9.	SreeDevi Book House Madras	-	1	4	20	5	3	-	33	8
10.	Kairali Children's Book Trust, Kottayam	-	-	23	4	1	2	-	30	9
11.	Janatha, Kochi	-	-	3	1	-	17	-	21	10
12.	Others	36	98	225	117	51	164	13	704	
	Total	74	173	402	286	142	289	24	1390	

constitute only 1.87 percent. In other words, compilations are few and far between in Malayalam. Books with corporate bodies as authors could be noted in Education, History, Political Science and Sociology. But their share is less than 1%. Books which do not specify the names of authors in the title pages appear in all subject fields. In some cases only the publisher's name is available in the title page.

4.1.5.2 Productivity of authors

There are certain writers in every subject who account for several books in their subject field. Analysis was made to identify the number of authors who had written more than one book. It was found that the number of books written by them range from one to twelve. In Law the maximum number of books written by an author was 2. In education upto 4 books were written. In geography it was six, in Economics it was seven and history nine. The maximum was in Political Science, a person with 12 books to his credit. The following table shows the number of books and the number of authors who contributed them.

It is clear from the table 6 that while 876 authors have written only one book each, the corresponding number of authors who have written two books is 119 and those with three books is 40. There is an inverse relationship between the number of books and the number of authors. Lotka has stated that the fraction of research workers in a given field who publish x papers is proportional to –n* where x is the number of papers and n is the index number. In chemistry the factor was 1/x2. In otherwords, Lotka found that the productivity of scientists conformed to Inverse Square Law such that for every 100 authors contributing one article, 25 will contribute 2 article, 25 will contribute 2 articles, 11 will contribute 3 and 6 will contribute 4.

In the equation $\frac{1}{x}$ n, the value of the index number n was given by Lotka as 2. This value was found to be applicable to chemistry. Here, if the value of n is considered to be 3, then Lotka's Law is found to be applicable, at least in the case of authors with 1, 2 and 3 books

Number of authors with one book	=	876
Number of authors with two books	=	119 (actual)
According to Lotka it should be $\frac{876}{2^3}$	=	109.5
Number of authors with three books	=	40

According to Lotka it should be $\frac{876}{3^3} = 32.4$

4.1.6 Publishers of Social Science Books

The entries of books were analysed to locate major publishers of social science books in Malayalam. For this purpose, publishers who had published ten and more books during the last 13 years were taken. The table 7 reveals the foremost publishers who publish social science books in Malayalam. Only the publishers who came upto the tenth rank in publishing social science books in Malayalam are included in the list.

Chintha Publishers, Thiruvananthapuram, the official publishing organ of the Communist Party of India (Marxist), tops the publishers with the highest number of books. This publisher alone is responsible for 10.36 percent of the total publications. Bhasha Institute (State Institute of Languages), Thiruvananthapuram, an autonomous body of the government of Kerala endowed with the responsibility of bringing out books suitable for higher education in Malayalam, comes next with 105 books. D.C. Books, Kottayam, the biggest private publisher in Malayalam, the Sahitya Pravarthaka Co-operative Society (SPCS), Kottayam, a co-operative society of writers in Malayalam and Prabhatham Publishers, Thiruvananthapuram occupy the third, fourth and fifth positions respectively. They are the foremost publishers of Malayalam books in social science. These 5 publishers account for 35.45 percent of the total. The rest of the 6 publishers contribute 13.88 percent.

4.1.6.1 Government Publications

The analysis showed that government agencies account for 9.28 percent of the total social science books. Table 8 shows the government/government sponsored institutions along with the number of books.

Table-8 Publications of Government/Government sponsored Institutions

Sl. No.	Institutions	No. of Books
1	Bhasha Institute	105
2	Department of Cultural Publications, Thiruvananthapuram	2
3	Kerala Sahitya Academy, Trissur	16

4	Publications Division, Govt. of India New Delhi	4
5	Public Relations Dept., Thiruvananthapuram	1
6	State Institute of Children's Literature, Thiruvananthapuram	1
	Total	129

As can be seen the Bhasha Institute is the foremost government publisher who brings out 105 books in social sciences.

4.1.6.2 Output of Scholarly Publishers

The term scholarly publishers usually refer to Universities, other academic institutions, professional bodies, research institutions etc. The analysis showed that the contribution of these institutions are sketchy. Except Calicut University which has brought out 3 books, no other University has published any book in Malayalam during the period under study. Malayalama is yet to receive any book published by research institutions in the field. The only professional association which had made any contribution in Malayalam was Kerala History Association, Ernakulam. They had broughout 5 books during the period 1981-1993. Since the contributions of Kerala Sahitya Academy and Bhasha Institute have already been considered as government publications, they are not included in the category of scholarly publications.

4.1.6.3 Author as Publisher

A notable feature of the book publishing in Malayalam is that some of the books are published by the authors themselves. Out of the 1390 books 123 books are published by the authors themselves. The following table indicate the number of books published by authors themselves in each subject.

Table-9 Author as Publisher in each subject

Subjects	Author as Publisher	Percentage
Education	9	7.32
Geography	42	34.15
History	30	24.39
Political Science	11	8.94
Economics	11	8.94

Sociology	19	15.45
Law	1	0.81
Total	123	100.00

The tendency to publish books by themselves is more prominent in Geography. Out of the 123 books published by the authors 42 are in that subject. Most of them are travelogues. In History there are 30 books and in Sociology it is 19. It is the least in the case of Law.

4.1.6.4 Geographical Location of Publishers

A district-wise analysis of publishers reveal that majority of the books are published from Thiruvananthapuram and Kottayam. 448 books (32.23%) are published from Thiruvananthapuram and 438 books (31.51%) from Kottayam. It has to be remembered in this context that Kottayam is hailed as the cradle of journalism with the two oldest dailies viz Deepika and Malayala Manorama stationed there. Ernakulam occupies the third place with 110 books (7.91%) and Trissur and Kozhikode occupy the fourth and fifth positions with 105 books (7.55%) and 85 books (6.12%) respectively. The rest of the district in Kerala account for only 5.76% of the total number of books.

A district-wise contribution of publishers who published Malayalam books in social sciences after 1980 is given below. It is arranged according to rank order.

Table-10 District-wise contribution of Publishers

Districts	No. of Books	Rank
Thiruvananthapuram	448	1
Kottayam	438	2
Ernakulam	110	3
Trissur	105	4
Kozhikode	85	5
Kannur	24	6
Kollam	22	7
Alappuzha	12	8
Palakkad	8	9
Pathanamthitta	7	10
Malappuram	3	11
Idukki	2	12
Vynad	2	12
Kasargode	-	-

It was seen that Malayalam books are published from outside the state also. Altogether 105 books were found to be published from other states. This comes to 8.92% of the total output during the period 1981-1993. Major centres of publication outside Kerala are given in the following table.

Table-11 Malayalam publishing outside Kerala

Sl. No.	Place of Publication	No. of Books
1	Madras	56
2	New Delhi	45
3	Bangalore	3
4	Bombay	1
	Total	105

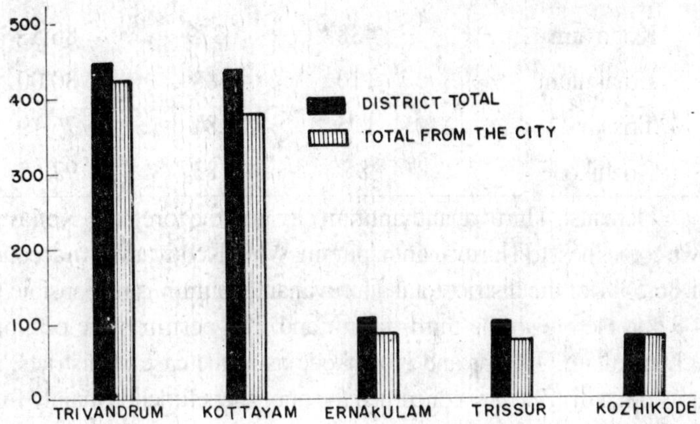

Fig. 3

Madras has contributed the highest number of books outside Kerala. It may be due to the fact that one of the oldest Malayalam printing press was established in Madras state and it still functions as one of the notable printing press in Malayalam. Madras University was the first university outside Kerala to start a full-fledged Malayalam Department and the Department has a good publication programme.

Malayalam books are printed and published from other countries too. Moscow accounts for the publication of 19 social science books during the period.

4.1.6.5 Important cities/towns that publish sizeable number of books in Kerala

Within each district, an analysis of the publishers was made to spotout the important cities/towns where there is a concentration of publishers.

Table-12 Important cities//towns that publish sizeable number of books in Kerala

Cities/Towns	District Total	Total from the city	Percentage
Thiruvananthapuram	448	425	94.87
Kottayam	438	379	86.53
Ernakulum	110	88	80.00
Trissur	105	80	76.19
Kozhikode	85	83	97.65

Here also Thiruvananthapuram city is in the forefront. Kottayam town comes next to Thiruvananthapuram. While Kottayam town accounts for 86.53% of the district total Thiruvananthapuram is responsible for 94.87%. Here also the third, fourth and fifth positions are occupied by Ernakulam, Trissur and Kozhikode as in the case of districts. As far as other districts are concerned the publishers function mainly from the district headquarters.

4.1.7 Various editions of Social Science Books

Many books in Social Sciences have more than one edition, which is an indication of the popularity of the books. The number of books that came out in various editions in different social science subjects is given below.

Table-13 Edition of Social Science Books

Subjects	1st	2nd	3rd	4th	5th	6th	7th	8th	9th	Total
Education	57	10	4	-	1	-	-	1	1	74
Geography	171	2	-	-	-	-	-	-	-	173
History	380	17	1	2	1	1	-	-	-	402
Political Science	276	5	1	1	1	2	-	-	-	286
Economics	133	7	2	-	-	-	-	-	-	142
Sociology	282	4	3	-	-	-	-	-	-	289
Law	22	-	2	-	-	-	-	-	-	24
Total	1321	45	13	3	3	3	-	1	1	1390
Percent--age	95.03	3.24	0.93	0.22	0.22	0.22	-	0.07	0.07	100

From the table it can be noted that the first edition constitute 95.03% of the total number of books. As can be expected the number of books decreases with subsequent editions. While 22.97% of books in Education have run into more than one editions the corresponding figures are 1.16%; 5.47%; 3.5%; 5.39; 2.42% and 8.33% for Geography, History, Political Science, Economics, Sociology and Law. It is clear from this table that the books in Education have more demand in the market. This may be due to the scarcity of books in that subjects. Subsequent editions are the least (1.16%) in Geography. The reason is that a good number of them are catalogues which are not revised. Education also comes first as the subject in which the highest number of editions have comeout. There are two books-one having eight editions and the other nine editions. The approximate proportion of social science books in general running into first, second and third editions is 102:3.5:1. In education, the proportion is 14:2.5:1 which is definitely higher.

4.1.8 *Translations*

Translations constitute an important part of Malayalam books in social sciences. Since they formed a sizeable category, subject-wise analysis of translations was carried out.

Table 14 provides a detailed account of the number of translations in each subject.

Table-14 Share of translations in each subject

Subjects	Total Books	Translations	Percentage
Education	74	9	12.16
Geography	173	8	4.62
History	402	74	18.41
Political Science	286	111	38.81
Economics	142	30	2.11
Sociology	289	36	12.46
Law	24	- -	- -
Total	1390	268	19.28

Translated works are more in Political Science where they form 38.81 percent of the total. History (18.41%), Sociology (12.46%) and Education (12.16%) occupy the next three positions. In Geography and Economics translations are less. In Law, all books are original. Books in various foreign languages such as English, French, Spanish, Russian, Iranian and Vietnamese are translated into Malayalam. Almost 50% of the translations are from English. Books in Indian languages especially Bengali are also being translated into Malayalam. Translations were found to be carried out by individual translators rather than translation agencies. Since many books did not contain information about the language from which the translations were made, an attempt to rank the languages on the basis of the translations was not made.

4.1.9 Price and Page Correlation

It is an established fact that the price of a book is decided on the basis of the number of pages, assuming that there is not much difference in the quality of the paper used. It is a truism that the price of the book has increase considerably over the years. In order to find out the rate of increase, the average prices of books in different ranges were calculated at different periods.

The average prices of social science books at 5 year intervals are taken except for 1993. The following table shows the correlation of pages with price at 5 year intervals.

Table 15 Correlation of pages with prices

Pages	1981	1986	1991	1993
1 - 100	6.00	9.00	15.00	20.00
101-200	12.00	19.00	25.00	31.00
201-300	18.00	27.00	35.00	55.00
301-400	20.00	38.00	45.00	67.00
401-500	21.00	48.00	50.00	100.00
501-600	22.00	70.00	80.00	100.00
601-700	30.00	40.00	85.00	125.00

It can be inferred that price almost double at ten year intervals and 50% increase at 5 year intervals. There are certain exceptions also. Between 1981 and 1986 90% increase is seen for 301-400 pages and 150% increase for 401-500 pages. The highest rate of increase is for 501-600 page books between the range of 5 years. The percentage of increase is 228. The average cost of 100 pages in 1981 was Rs. 5.00; Rs. 9.00 in 1986; Rs. 12.00 in 1991 and Rs. 18.00 in 1993. It shows that there is an increase of 80% between 1981-1986; 60% in 1986-1991 and 300 percent between 1991-1993. During the 90's prices are shooting up at a faster rate.

The above data is analysed using correlation coefficient (r)

$$r = \frac{\text{cov}(xy)}{sx \,.\, sy}$$

where $\text{cov}(xy) = \frac{\Sigma xy - \bar{x}\,\bar{y}}{n}$

and $sx = \sqrt{\frac{\Sigma x^2}{n} - x^{-2}}$

(sx = standard deviation of x)

$sy = \sqrt{\frac{\Sigma y^2}{n} - y^{-2}}$

(sy = standard deviation of y)

The correlation coefficient between number of pages and price of books in the year 1981 is found to be 0.957. This indicates that the price of books is highly correlated to the number of pages. In the year 1986 the correlation coefficient is found to be 0.83; in 1991 it is 0.978; in 1993 is 0.988. The above analysis shows that the price of books is highly correlated to the number of pages in the year 1993.

4.1.10 *Comparison of prices between the 5 biggest publishers*

A comparative study of the prices of biggest publishers namely Chintha, Bhasha Institute, D.C. Books, Sahitya Pravarthaka Sahakarana snagham (S.P.C.S.) and Prabhatham for the period 1980-1993 has been carried out. The table shown below provides details regarding the range of prices per page.

Table-16 Comparison of Price structure of the top 5 publishers

Publishers	Average price for 100 pages (in Rs.)	Average price for200 pages	Average price for300 pages	Average price for400 pages	Average price for500 pages	Average price per100 pages
Chintha	13.50	15.75	35.00	38.00	50.00	10.50
Bhasha Institute	10.00	16.25	17.00	18.00	27.00	6.75
D.C. Books	9.00	20.50	36.00	48.00	65.00	11.50
S P C S	13.50	29.00	33.50	50.00	102.50	14.50
Prabhatham	6.75	27.50	29.25	30.00	40.00	9.25

With regard to the cost of books those published by the Bhasha Institute have the lowest cost and those of SPCS have the highest. The difference in average price between the two is 53.45%. This indicates that the price of Bhasha Institute. Among the private publishers, the average cost for 100 pages is least for Prabhatham, Thiruvananthapuram. Chintha comes next and D.C. Books occupy the third position. SPCS, though a co-operative society of writers, charges the highest rate among publishers. But a comparison of the actual publications reveal the reasons for the low cost.

The comparative study of the prices between the 5 biggest publishers namely Chintha, Bhasha Institute, D C Books, SPCS and Prabhatham is analysed using correlation coefficient. The correlation coefficient between the number of pages and average price of Chintha Publishers is found to be 0.970 indicating high positive correlation. For Bhasha Institute it is 0.9279 and for D C Books it is 0.998. For SPCS publishers it is 0.92 and for Prabhatham it is 0.8989. Thus it can be concluded that correlation is more or less proportionate for the aforesaid publishers.

4.1.10.1 Comparison of price between Government and Private publishers.

The table given below shows the average cost of books of private publishers and government publishers

Table-17 Comparison of price between Government and Private Publishers

Publishers	Average cost of Books (in Rs.)
Private publishers	11.50
Bhasha Institute	6.75
Other Government publishers	11.50

From the table it could be seen that the average price of the books of private publishers per hundred pages was Rs. 11.50. This was obtained by taking the average cost of 100 pages charged by Chintha, D C Books, Prabhatham and SPCS. The average price of other government publishers was 11.50 and that of Bhasha Institute Rs. 6.75. When the average of these two figures are taken it becomes Rs. 9.00 per 100 pages. This shows that the price charged by private publishers is 21.73% higher than that of government publishers. It can also be seen that, there is not much difference between the price charged by private publishers and other government publishers except Bhasha Institute.

5. CONCLUSION

There were altogether 1390 Malayalam books in social sciences published from 1981-1993. But there is an unbalanced growth of books in different subjects. While History books constitute more than a quarter, Education and Law lag behind in terms of the number. Within each subject also there are certain areas where books are yet to appear in Malayalam.

The study reveals that private publishers are keen on bringing out social science books in Malayalam. They publish 90.92 percent of the total. The lack of interest of universities, other academic and research institutions and professional bodies in the publication of books in Malayalam was evident from the study.

It is also brought to light the need for establishing translation agencies to translate good social science books from other languages to Malayalam. At present translations are brought out by individuals alone. In the field of literature Kerala Sahitya Academy takes the lead in getting works translated into Malayalam from other languages and vice versa. But there is no such set up to take care of social science books. The role of National Book Trust (NBT) was found to be nil in this regard.

The study also brings out the absence of effective bibliographic control of Malayalam books in general and social science books in particular. There are some journals in Malayalam which enlists the literary works in Malayalam. But non-literary works are left in the lurch. Ofcourse, Malayala Granrt hasoochi and Indian National Bibliography are there, but they lag behind very much. It is therefore, high time that individuals and institutions take initiative to bringout comprehensive bibliographies in Malayalam in the social science subject.

REFERENCES:

1. Agrawal, SP : National information resources for social sciences in India: Mobilisation, dispersal and International enhancement. New Delhi, Concept PNB, 1992.
2. Agrawal, SP : Social Science information and documentation, New Delhi, Concept PNB, 1987.
3. Binwal , JC, et al. : Social Science Information: problems and prospects, New Delhi, Vikas, 1990.
4. Govi, KM : Malayala Granthasoochi. V. 1-6 Trissur, Kerala Sahitya Academy, 1973-1993.
5. Govi, KM : Malayala Pusthakangal enpathukatil (Malayalam books in the eighties) Mathrubhoomi Daily 6th June, 1993.
6. Indian National Bibliography Cumulative Volumes 1980-90. Calcutta, Central Reference Library, 1993.
7. Ramareddy, E : Social Science information: Some Indian Sources, New Delhi. Affiliated East West Press, 1985.
8. Ranganathan, SR : Social Sciences: Emergence and Scope. Library Science with a slaut to Doc. 9 (2) 1972.
9. Webb, William H, et al : Sources of information in the Social Sciences ed.3 New Delhi, Arnold Heineman, 1986.
10. Vyas, SD : Social Science information in India : Efforts towards bibliographical Control New Delhi, Concept PNB, 1992.

6

BIBLIOMETRIC ANALYSIS OF SCIENCE JOURNALS PUBLISHED FROM INDIA

DR. JALAJA V

0. INTRODUCTION

The exponential growth of literature and rapid development of libraries generated several evalutionary studies about the effectiveness and efficiency of information services. These studies led to the identification and application of appropriate quantitative measuring techniques known as bibliometrics. Libraries and Information managers all over the world began to use bibliometric techniques in their day to day administration. These bibliometric studies throw light on the pattern of growth of literature, inter-relationship among different branches of knowledge, productivity and influence of authors, pattern of collection build up, their use etc. Day by day bibliometric is attaining inter disciplinary character and sophistication.

Reference materials are required for up-to-date information in various fields due to the explosion of knowledge and with the rapid changes noticed in all domains of human activity. But efficiency cannot be ensured unless this service is planned and executed in accordance with precise techniques of modern library and information science. Bibliometric is one of such modern development in library science that help library personnel to design accurate strategies associated with their work.

1. RELEVANCE OF THE STUDY

Rising journal subscription costs, shrinking budget allocation, and lack of space for current library holdings have resulted in number of 'document use' studies. Librarians have been trying to determine library collection development policies which will make optimum use of money and space available while meeting the current and future needs of the library users. The number of new titles of journals has grown at an alarming pace while subscription prices of established journals have been rising at a rate much higher than the rate of inflation. Libraries have been forced to cancel certain titles in order to keep up with increasing costs, as well as not purchasing new titles and making room for the current ones.

Escalation in the cost of scientific publications, the lack of adequate library budgets and the different fiscal years ahead, have all created problems both to the users of information as well as to the providers and organizers of information. The agencies which have to organize and provide information resources and services have to understand the requirements of users of information as well as the characteristics of the subject literature, inorder to develop a need - based information resources and services to meet the information requirements of scientists working in various fields. Therefore, a clear understanding of the characteristics of subject literature is one of the prerequisites and has paramount importance in planning and designing of information systems and service in a particular field. The characteristics of subject literature such as patterns of growth and scattering productivity, dissemination, distribution, communications, obsolscence patterns, utilization and consumption etc. have all been found very useful in planning, developing and organizing information resources and services and also in taking meaningful decisions in solving some of the problems of libraries, documentation and information centres. Hence, in this study, which is based on Indian Science Abstract literature analysis of references have been undertaken to study the characteristics of science literature.

2. OBJECTIVES OF THE STUDY

The specific objectives of this study are the following:

1. To identify a list of core Indian journals in Mathematics, Astronomy and Astrophysics, Physics, Chemistry, Geology, Biology, Botany, Zoology, Medical Sciences, Engineering, Agriculture, Chemical Technology and Manufacture and to study the scattering pattern of literature of these subjects in different journals.

2. To determine the authorship pattern in the above subjects and to find out whether there is any difference between this study and similar studies earlier.

3. To identify the trends in research contributions in pure sciences

as well as in applied sciences in India.

4. To find out whether foreign authors are interested in publishing their articles in Indian science journals.

5. To determine the extent of research contributions in pure and applied sciences form Kerala state.

6. To determine the institutional affiliation of contributions in pure and applied sciences.

7. To obtain form dispersion of references in the above thirteen subjects.

8. To determine the yearwise productivity in the thirteen subjects.

3. METHODOLOGY

This study is based on the abstracts published in the secondary journal, Indian Science Abstracts (ISA), for a period of five years ie. from 1984 to 1988. During this period of study ISA had covered 21 subjects in all with 2753 journals. These subjects which have been chosen for detailed study. So thirteen main subjects are selected such as Mathematics, Astronomy and Astrophysics, Physics, Chemistry, Geology, Biology, Botany, Zoology, Medical Sciences, Engineering, Agriculture, Chemical Technology and Manufacture. Survey and document analysis are the main techniques used for collecting data for this study. In addition to this document analysis, preparation of frequency distributions, calculations of percentages, graphical representations and statistical analysis are also used.

4. ANALYSIS

The variables analysed are :

1. The journals referred in the abstract
2. Number of times a particular journal was referred
3. Different forms of publications
4. Nature of authorship
5. Extent of contributions of foreign authors in Indian journals
6. The status of contributions form the Kerala state
7. Yearwise productivity

Core journals are identified in the above mentioned thirteen subjects on the basis of the frequency of repetitions of references. According to the number of references ten core journals were identified in each subjects. After that it was tested to know whether Bradford Law of scattering is followed in each of the thirteen subjects or not.

Inorder to assess the trend of author productivity analysis of the reference with respect to authors has been done. This data was compared with findings of another similar study. The data was further analysed to know the predominant agency and its contributions were determined.

Scanning of the documents were done inorder to determine the number and percentage of foreign authors contributions in Indian journals.

The documents were thoroughly examined to reveal the percentages of journal articles, theses, patents and standards in science contributed from the Kerala state during the study period by means of frequency count. In addition to all these, form analysis and yearwise productivity were also ascertained.

5. FINDINGS

The following are the major findings that are drawn form this study :

1. Scattering of science periodicals follows Bradford's law of scattering in all the 13 subjects studied. Core journals, those which are in the nucleus, of each of the above subjects were found out. Among these 13 subjects highest number of journals in the nucleus is 5 and Mathematics and Botany had 5 core journals each. Biology and Zoology had 4 core journals each. In the three subjects - physics, Engineering and Manufacture 3 core journals were found in the nucleus. Astronomy and Astrophysics, Chemistry, Geology and Medical Sciences had 2 core journals each. In chemical Technology there were only one core journal in the nucleus.

In addition to core journals, prominent journals in the above subjects were also found out. In Mathematics there are 40 such journals. These prominent journals had a coverage of 94.93 percent of the total references. Botany secured the second position with 16 prominent journals with a coverage of 85.64 per cent references. Manufacture attained the third position with 14 prominent journals having a coverage of 76.70 percent references. Chemical Technology had the last position with 3 prominent journals with the coverage of 28.01 percent references. Thus, it emerged that among the 2243 journals 118 prominent journals covered more than 50 percent of the total references.

Apart from these, values of 'b' for these different subjects were found out. Among these subjects Zoology attained the maximum value ie. 0.7715 and the minimum value was secured by manufacture ie. 0.4644. For the remaining 11 subjects the value of 'b' varies between 0.7222 and 0.5227. The equation R(n) = and value of 'b' was not a constant for 5 years. Thus, the time factor could not be extended form 3 years to 5 years.

2. Joint authorship is dominant in all the subjects except Engineering. Among the subjects, Chemistry had the highest percentage of joint authored references. Compared to other subjects, difference between single authored and joint authored references were relatively low in Mathematics and Astronomy and Astrophysics. In Chemistry it is the team research which dominated the research and development activities. But in the case of Mathematics it is the individual's mind. By using X^2 test the difference between these two in two years ie., 1984 and 1988 were found, $X^2 = 630.988$ with degree of freedom 1, which is found to be significant differences.

On comparing the findings of this study with a similar study related to 1965-83, it was found that the tendency towards joint authorship was on an increase This indicated that researchers normally prefer co-operative work. The other factors such as nature of the project, magnitude if instrumentation, popularity and the rate of expansion of the area of science may also affect this tendency.

3. Analysis of references reveals that most of the research works in Pure Science except in Astronomy and Astrophysics and Geology were being pursued in Universities and other higher educational Institutions. Research institutes/laboratories were more active in the field of applied sciences. The relation between the references form Universities and Colleges on the side and the research institutes/laboratories on the other side in 1984

and 1988 were found out by using X^2 test. The value of X^2 = 2184.27 at degree of freedom 1 was found to be significant at 0/001 level.

4. Geographical analysis of references revealed that during the period of study contributions form abroad were more in pure sciences than in applied sciences. Research in applied science required good laboratory facilities and other modern infrastructure. If we compare these facilities in developed countries to that of India, India stood behind. And research in developed countries in applied sciences is more advanced than that of India. This may be attributed as a reason for less number of references in applied sciences.

In Physics, Chemistry, Geology and Medical Sciences, contributions from abroad were increasing. The extent of contributions from abroad in pure science and applied science in 1984 and 1988 were compared by using X test. The value of X^2 = 73.859 with degree of freedom of 1 and the difference is significant at 0.001 level.

Geographical analysis of contributions from abroad and other Indian states showed slight decreasing tendency. But when these data were tested by means of X^2 test it is found that there is no significant difference in contributions from abroad and contributions form other Indian states during this period of study. Analysis of the contributions form Kerala revealed an increasing tendency. This data was also statistically tested by applying X^2 text and it was found that there is significant difference in contribution form Kerala during this period of study.

5. Status analysis of references from Kerala showed that it was nominal in both pure sciences and applied sciences with regard to contributions from Kerala, in pure sciences Geology secured the highest per cent of references and in applied science Manufacture stood first.

6. Form analysis of the references showed that there were four forms such as journal articles (90.28 per cent), patents (1.78 percent), standards 91.75 percent) and theses (0.19 percent). This shows the dominance of journals among forms such as

patent, standards and theses.

7. Among the total number (991529) journal articles 39.33 percent were in pure sciences and 60.67 per cent were in applied sciences. Agriculture had maximum number of articles.

8. During the period of study there were references of 1690 patents. Among these, 92.33 per cent were in applied sciences.

9. The total references of standards were 1661. Engineering with 486 standards stood first and Mathematics with 2 standards was the last. Here also applied sciences was dominating with 63.37 per cent of standards.

10. During the period of study 180 theses were referred in eleven subjects. Theses were dominating in pure sciences with 70.77 per cent references. Chemistry with 31 theses top the list and Manufacture with 5 theses was at the bottom.

11. The productivity of subjects in general shows a trend of exponential increase. Agriculture was found to the most productive subject followed by Medical Sciences. Among the 13 subjects only Chemical Technology and Manufacture showed a slight decrease in references in 1988. The references in applied sciences is more than that of pure sciences. The X^2 test indicates that there is significant difference in the number of publications under these two categories.

6. SUGGESTIONS FOR IMPLEMENTATION

1. As a result of enormous growth of knowledge the number of publications has also increased. Due to the shortage of fund most of the libraries are compelled to restrict the subscription of periodicals and purchase of other reading materials. In order to get maximum benefits the libraries can limit the procurement by subscribing to core journals in the main subjects. Materials in developing subjects can be acquired by means of other methods of procurement. It must be ensured by all the libraries in a geographical area to see that at least one of the core journals is available in any one of the libraries in a locality. This forces to establish a city library network, regional networks, national networks, access to international databases, union catalogue

and resource sharing. Selection of core journals in this regard has to be made using appropriate bibliometric techniques.

2. The dominance of joint authored papers indicated to the emergence of team work in research activities. Hence team research must be encouraged.

3. In the field of research in applied sciences, Universities and colleges must be given more encouragement.

4. This study found that foreigner's interest in publishing papers in Indian journals is increasing. So necessary steps should be taken by the Indian publishers to encourage this trend, so that more and more foreign libraries will subscribe to Indian journals. Necessary detailed information has to be supplied to foreign institutions, regarding the scope and coverage of Indian journals and other publications.

5. The number of theses, patents and standards will definitely be more than those referred in Indian Science Abstracts. According to ISA due to the lack of proper information many of them could not be included. Hence attempts must be made to have an infrastructure to get up to date information about all forms of Indian publications in science and technology.

6. On the basis of this type of studies the collections can be subdivided into primary, secondary, and tertiary sources or into stored having frequent, occasional or only rare access.

This will facilities more convenient and quick references of those materials that most wanted by researchers and other academic workers.

7. The allocation of money for various journals, patents and standards for the different subjects should be regulated on the basis of accurate findings regarding the frequencies of references made, arrived at through bibliometric studies. A scheme for sharing of journals, patents and standards between a central reference point and satellite departmental or regional collection may be developed on the basis of the findings.

6. SUGGESTIONS FOR FURTHER RESEARCH

1. Research studies aiming at identifying core journals, prominent journals and insignificant journals in area other than science and technology should be taken up.

2. The data for the present study was gathered soley from Indian Science Abstracts, more comprehensive studies on the basis of data obtained from other sources should be undertaken to supplement the findings arrived at in this study.

3. In order to verify whether some of the deviations form the established laws of bibliometrics noticed in this study, similar studies need be conducted.

4. Studies giving more stress to the bibliometircs of patents, standards and theses need be carried out.

REFERENCES:

1. Arjun Lal. "Ranking of periodicals in the field of social Sciences". Annals of library science and documentation 37(2,1990), pp.67-73.
2. Banik, B.D. and Bandyopadhaya, S. "Recent growth of periodicals on glass and ceramics - Their rank". IASLIC Bulletin 30 (1,1981), pp 23-30.
3. Bavakutty, M and Taher, Mohamed. "Flow of Islamic Information through journal articles : A bibliometric analysis". Lucknow Librarian 22 (3-4, 1991), pp 75-80.
4. Brookes, B.C. "Theory of the Bradford's Law". Journal of Documentation 33 (4, 1997), pp. 180-209.
5. Deniel, R S. "Bibliometrics and scholarly impact". American Psychologist 34 (8, 1979) pp. 725-728.
6. Elisha Raju, E and Kavitha. "Core journals in synthetic organic chemistry." Annals ot library science and documentation 32 (3-4, 1985) pp. 115-121.
7. Guha, B. Documentation and information services techniques and systems, Ed.2. Calcutta, World Press, 1984.
8. Gupta, D K. "Application of Bradford's law of citation data of Ethioplan Medical Journal". Annals of library science and documentation. 38(3,1991) pp. 183-188.
9. Lalitha Kumari and Sengupta, I N. "Bibliometric evaluation of Bio Chemical research in India, Part II. "Indian journal of information, library and society 3 (102, 1990) pp. 16-28.
10. Maheswarappa, B S. "Characteristics of Phytopathology literature: A

bibliometric study based on primary phytopathology periodicals (1966-85). Ph.D. Thesis, Gulburaga University, Department of Library Sciences, 1982.

11. "Research collaboration in different disciplines of applied science in India (1965-83): A bibliometric study". IASLIC BULLETIN 32 (3, 1987), pp.105-114.

12. Narendra Kumar. "Literature contribution in Journal of Environmental Biology : A bibliometric study". Annals of library science and documentation 33 (1-2, 1988) pp. 31-40.

13. Ravichandra Rao, I K. "Bibliometrics: Its theory and practice." Proceedings of the DRTC Refresher Seminar, 1981, Paper AAPT, 16.

14. "Quantitative Methods for library and information Science, New Delhi, Wiley Eastern, 1983.

15. Sangam, S. L. A citation analysis of doctoral dissertations in social science accepted by the Karnataka University, Dharward during 1984 to 1982 (Thesis) Gulburga University Department of library science, 1985.

16. Sen, B K. and Narendra Kumar "Indian contribution in bibliometrics. 1958-1984: A review "Annals of library science and documentation 33, (3,1986) pp.85-103.

17. Sengupta, I N. "Bibliometrics: A birds Eye View". IASLIC Bulletin 30 (1, 1985), pp. 168.

7

CONTRIBUTION OF KERALA TO INDIAN SCIENCE: A BIBLIOMETRIC ANALYSIS

K G SUDHIER

0. INTRODUCTION

Science and scientific research have obviously been growing at a faster rate during recent years. Every nation considers scientific and technical development as the basis for its overall progress. The manpower engaged in scientific research also show considerable increase in recent years. Similar is the case with the number of institutions engaged in scientific and technical education and research.

Scientific research assessment is a complex process which can be simplified by measuring it indirectly using bibliometric indicators based on the estimation of the output of science, which implicitly indicates productivity. For decades publication counting has been outcome of research. But the major limitation of this method is that it can measure only research activities that result in written knowledge which is scattered by the established communication channels like journals, books, patents etc.

In spite of the limitation of this method, it was considered worth while to study the nature and other characteristic features of scientific research in Kerala as evidenced by the scientific publications emanating from the state.

1. THE PROBLEM

As far as Kerala is concerned, though the State was in the forefront of education, the same thing could not be claimed in scientific investigations. It is generally accepted both by the educational administrators and the general public that though the achievements of Kerala in the field of general education are incomparable, the state has miserably failed in the diversification of courses at the higher level. In the field of academic research also, it is believed, the state has fallen short of expectations. This has prompted the investigator to verify the validity of the statements by analysing the publications emanating from the State.

2. OBJECTIVES

The main aim of the study is to investigate into the characteristic features of the scientific research in Kerala by analysing the contributions of scientists and deduce valid conclusions.

In order to achieve this, the following objectives were set:

i) To examine the growth of S & T papers in Kerala during the last fifteen years.

ii) To identify the areas of research in the State and the relative contributions in each area.

iii) To study the authorship pattern.

iv) To find out the relative contributions of different types of institutions in the State.

v) To examine the scientific productivity of different districts and thereby establish the strong holds of scientific research in the State.

3. METHODOLOGY

Source of Data

In order to create a bibliographic database of publications in science and technology in Kerala, several possible sources were considered. As far as Indian scientific publications are concerned, Indian Science Abstracts (ISA) is the most comprehensive secondary source. Hence it was decided to locate the contributions of the researchers in Kerala from this publication.

ISA is a prestigious publication of the Indian National Scientific documentation Centre (INSDOC), New Delhi. It started in 1965 with a view to record comprehensively the output of the Indian scientific documents and also as an effective feeder service to the International Abstracting Service. The frequency of ISA was changed from monthly to semi-monthly in 1985.

The ISA reports works published in India as well as work done in India but published in foreign periodicals by Indian scientists. It covers original (research) articles including short communications, case reports, review and informative articles published in scientific and technical periodicals, conference proceedings, as well as patents and theses. About 1,000 records are being documented in each semi-monthly issue.

Period covered

The period chosen for the study was from 1979 to 1994. This period of 15 years was judged long enough to reveal the major trends in the scientific output of the State.

From the 29 volumes completed so far (vol. 30 of 1994 is incomplete), 4 sample volumes at five year intervals beginning from the year 1979 were taken for study. Due to some technical reasons the publication of 30th volume delayed inordinately and therefore only 14 issues upto August 1994 could be taken for analysis.

The four volumes of ISA were scanned to collect the necessary data and all the relevant details about the papers were recorded in 7.5 X 12.5 cm cards. The details in each entry included the main subject area, its subfields, name(s) of the contributor(s), address of the contributors, title of the paper, title of the journal in which it appeared, its bibliographic details, page numbers and number of citations.

Different aspects viz, author, subjects, number of papers, institutions, geographical distribution etc. were taken into account for the analysis and interpretation of the collected data.

4. NUMBER OF PAPERS

Scientific papers are the outcome of scientific research. Naturally the number of papers published on a subject will give an indication of the extent of active research that is being carried on the subject. Therefore, the total number of contributed articles were counted from different sample volumes and the results are indicated in Table 1.

During the five year period from 1979-84, while the total number of papers increased by 40.42%, the corresponding increase for papers from Kerala was more than 70%. In the next five years, ie from 1984-89, the increase in all India and Kerala percentages were 3.26% and 37.13% respectively. Similarly, for the next five years from 1989-94, the total number of papers increased by 12.34% and the corresponding increase of the Kerala's share was 19.80%.

A comparison of the number of papers in the 4 volumes shows that the total number of papers have increased 27.10% and the corresponding increase of papers from Kerala was 291.67%

A diagrammatic representation of the increase in the number of papers since 1979 is given in the Figure-1.

While the share of the science literature from Kerala was only 1.156% in 1979, its share rose to 2.652% after 15 years. The figures in the table indicate that in 15 years time the annual contributions have increased three fold, ie from 180 in 1979 to 525 in 1994.

Table-1 Year-wise distribution of Papers

	Year				
	Vol.15 1979	Vol.20 1984	Vol.25 1989	Vol.30 1994 (upto 16 August)	Total
Total No. of papers	15,573	21,867	22,580	19,580	79,813
Papers from Kerala	180	307	421	525	1,433
% of contributions	1.156	1.404	1.864	2.652	1.795

5. SUBJECT-WISE DISTRIBUTION

The titles of the papers were examined in order to ascertain their specific subject. For the analysis, major subject divisions from UDC Abridged Edition were used and the subject headings given by ISA were also helpful to determine the specific subjects.

The subject-wise break up of articles covered in the study is given the Table-2. In the year 1979, the majority of the contributions were from Agriculture and related fields (27.22%), followed by Medical Sciences (16.11%) and Zoology (14.44%).

In vol. 20, ie in the year 1984, Agriculture covered 26.71% of the total contributions. Engineering and Technology occupied 17.59% followed by Zoology (11.73%)

In 1989 (vol.25) Agriculture field covered 32.30% and Medical Sciences 14.96%. In 1994 (Vol.30) 32.38% of the contributors were from Agriculture followed by 12.38% from Medical Sciences.

Table-2 Subject-wise break up of the contributions

Subjects	1979 No. of papers	1979 % of papers	1984 No. of papers	1984 % of papers	1989 No. of papers	1989 % of papers	1994 No. of papers	1994 % of papers	Total	
1. Mathematics and Natural Sciences Mathematics	1	0.55	5	1.63	12	2.85	6	1.42	24	1.67
2. Physics & Mechanics	7	3.96	12	3.90	42	9.98	21	4.00	32	5.72
3. Chemistry, Crystallo-graphy, Minerology	12	6.67	25	6.14	21	4.99	32	5.09	90	5.28
4. Geology, Meteorology	5	2.78	24	7.81	21	4.99	28	5.33	78	5.44
5. Biology Anthropology	17	9.44	16	5.21	25	5.94	35	6.67	93	6.49

Subjects	1979		1984		1989		1994		Total	
	No. of papers	% of papers	No. of papers	% of papers	No. of papers	% of papers	No. of papers	% of papers		
6. Botany	11	6.11	14	4.56	31	7.36	57	10.86	113	7.88
7. Zoology	26	14.44	36	11.73	44	10.45	53	10.09	159	11.09
8. Applied Science Medical Science Health & Safety	29	16.11	18	5.35	63	14.96	55	12.38	175	12.21
9. Engineering & Technology, generally.	11	6.11	54	17.59	11	2.51	18	3.43	94	6.56
10. Agriculture Forestry, Stock freeding, Fisherness.	49	27.22	32	26.71	136	32.30	170	32.38	437	30.49
11. Chemical Industry, Chemical Technology,	6	3.33	16	5.22	11	2.61	20	3.80	53	3.70
12. Manufacturing Industries &	6	3.33	5	1.63	-	0.95	20	3.80	35	2.44
Total	180	100	307	100	421	100	525	100	1433	100

Table-3 Authorship pattern

No. of Authors	1979		1984		1989		1994		Total	
	No. of papers	% of papers	No. of papers	% of papers	No. of papers	% of papers	No. of papers	% of papers	No. of papers	% of papers
1. Single Author	48	26.77	77	25.08	70	16.63	129	24.57	324	22.61
2. Two Authors	65	36.11	122	39.74	196	46.55	192	36.57	575	40.12
3. Three Authors	46	25.55	64	20.85	91	21.61	109	20.76	310	21.63
4. Four Authors	17	9.55	31	10.10	38	9.03	52	9.90	138	9.63

No. of Authors	1979		1984		1989		1994		Total	
	No. of papers	% of papers	No. of papers	% of papers	No. of papers	% of papers	No. of papers	% of papers	No. of papers	% of papers
5. More than 4 Authors	4	2.22	13	4.23	26	6.17	43	8.19	86	6.00
Total	180	100	307	100	421	100	525	100	1433	100

It is clear from the table that from among the pure science subjects Zoology comes first with the largest number of contributions followed by Botany.

Among the Applied Science group Agriculture, Forestry and Fisheries had a definite edge over other subjects.

Judged from the research output, it can be safely stated that applied research, especially in the fields of Agriculture and Medicine is more predominant than pure research in Kerala.

6. AUTHORSHIP PATTERN

With a view to identify the extent of research conducted by individuals in collaboration with each other, the number of authors mentioned in the papers was counted and analysed. Table-3 gives the distribution of articles on the basis of the number of authors.

The single author contributions accounted 26.67% of the papers covered in 1979. In the same year the contributions by two authors amounted to 36.11% and the share of three author papers was 25.55%.

In vol. 20, ie in 1984, single authors contributed 25.08% and joint authors about 75%.

In the year 1989, single author contribution is definitely less than that of the previous years. Joint papers accounted for 83.37% of the total. Two author papers were responsible 46.55% of the total.

In 1994, two author papers occupied 36.57% of the total contribution. Joint author papers accounted for 75.43% of the total number of papers.

It has to be observed that two-author papers retained their supremacy in all the four years. A noticeable trend is the constant increase in the number of papers with more than 4 authors. While their share was only 2.22% in 1979, it was 8.19% in 1994.

7 CONTRIBUTION OF DIFFERENT TYPES OF INSTITUTIONS

Nowadays, scientific research is carried out by different types of organisations. They include academic institution like universities and colleges, industrial organisation, R & D centres and so on.

In orders to ascertain the contributions from the different types of organisation, the institutions, the institutional affiliation of the authors was examined and analysed. The results are given in Table 4.

For the convenience of the study, the institutions were grouped into 3 categories:

1) Universities and Colleges

2) Research institutions

3) Private and other institutions.

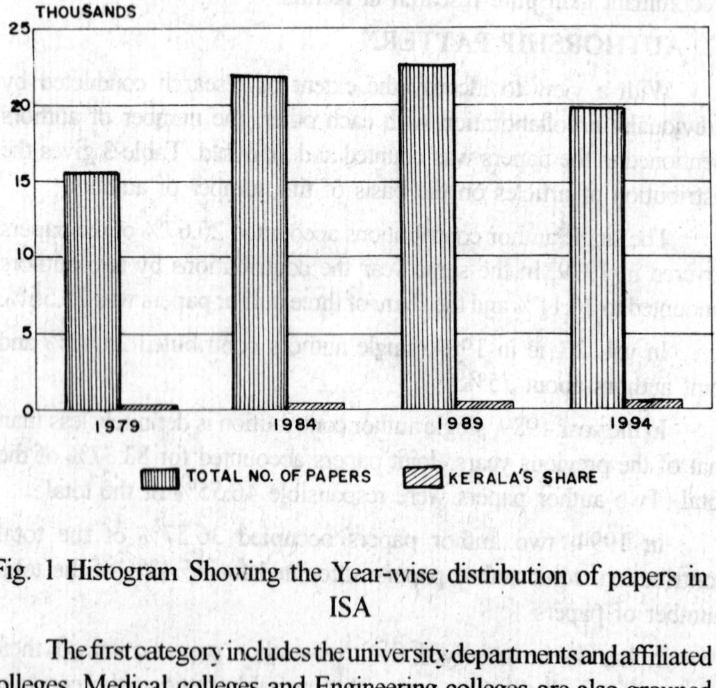

Fig. 1 Histogram Showing the Year-wise distribution of papers in ISA

The first category includes the university departments and affiliated colleges. Medical colleges and Engineering colleges are also grouped with colleges, since their contributions were found to be relatively small. Research institutions of central and state governments are clubbed together.

In the year 1979, the quantity of the publications from academic institutions was almost double of that from research institutions. In 1984, this trend was totally reversed. While academic bodies contributed 42.34% of the publications, the share of research institutions had gone upto 54.4%. This can be attributed to the large number of research institutions that have come up during 1970's under the auspicious of the state government.

In 1989, Universities and colleges again regained their supremacy with a share of 61.28% of the papers. The contributions of research institutions had come down to 35.39%.

In 1994, universities and colleges had contributed 57.71% and the share of research institutions was 35.43%.

Table-4 Institution-wise contribution of papers

Bodies	1979	1984	1989	1994	Total
1. Universities & Colleges	116 (64.44%)	130 (42.34%)	258 (61.28%)	303 (57.71%)	807 (56.31%)
2. Research Institutions	62 (34.44%)	167 (54.40%)	149 (35.39%)	186 (35.43%)	564 (39.36%)
3. Private & other Institutions	2 (1.11%)	10 (3.26%)	14 (3.32%)	36 (6.86%)	62 (4.33%)
Total	180 (100%)	307 (100%)	421 (100%)	525 (100%)	1433 (100%)

The striking revelation of the data is the decline in the contribution of the research institutions in the last decade. Figure-2 is a diagrammatic representation of the share of the papers among different institutions.

In order to analyse the performance of the universities in the state in the research front, the relevant entries were separated and subjected to detailed analysis. The results are tabulated in the Table 5. It is evident from further analysis of the data that research in Agriculture is more active in the state. It was also found that in the years 1979 and in 1984 University of Kerala stood first and Kerala Agricultural University stood second. In 1989, the Agricultural University had almost caught up with University of Kerala. In 1994, Agricultural University had overtaken other universities, and Kerala University was pushed to the second position.

From the Histogram showing total contribution of the Universities (Fig-3), it is seen that for the last 15 years there is a steady increase in the contributions only in the case of Kerala Agricultural University

8. GEOGRAPHICAL DISTRIBUTION OF THE CONTRIBUTORS

Details about the official address of the contributing authors were examined from the address given in the entries and the districts of their parent body were ascertained. The district-wise contributions were analysed and the results are given in the Table-6.

In 1979, out of the 180 papers from the state, 75(41.67%) were from Thiruvananthapuram district and 30 (16.67%) were from Ernakulam district. Trissur occupied the third position, Malappuram and Kottayam claimed the 4th and 5th position respectively.

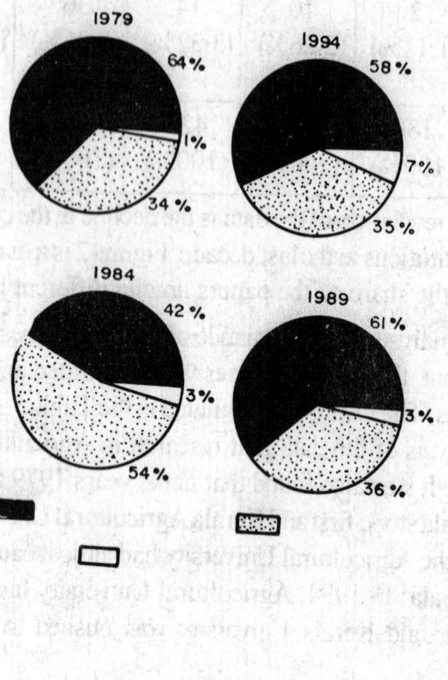

Fig. 2 Diagrammatic Representation of the share of the paper among different Institutions in different Years

Table-5 University-wise distribution of papers

Universities	1979		1984		1989		1994		Total	
No. of	% of Papers	No. of Papers	% of Papers	No. of Papers	% of Papers	No. of Papers	% of Papers	No. of Papers	No.	%
1. University of Kerala	52	44.83	65	50	77	29.84	86	28.38	280	34.70
2. Calicut University	25	21.55	18	13.64	52	20.15	45	14.85	140	17.35
3. M.G. University	- -	- -	- -	- -	21	8.14	18	5.94	39	4.83
4. Cochin University of Science & Technology	12	10.34	20	15.38	33	12.79	49	16.17	114	14.13
5. Kerala Agricultural University	27	23.27	27	20.77	75	29.07	105	34.65	234	29.00
Total	116	100	130	100	258	100	303	100	807	100

Table-6 District-wise distribution of contributors

Districts	1979		1984		1989		1994		Total	
No. of	% of papers	No. of papers	% of papers	No. of papers	% of papers	No. of papers	% of papers	No. of papers	% of	
1. Trivandrum	75	41.67	199	64.82	164	38.95	174	33.14	612	42.17
2. Kollam	2	1.11	2	0.65	2	0.47	18	3.34	24	1.67
3. Alappuzha	14	7.78	3	0.98	3	0.71	6	1.14	26	1.81
4. Pathanamthitta	- -	- -	- -	- -	2	0.47	1	0.19	3	0.21
5. Kottayam	17	9.44	5	1.62	25	5.94	48	9.14	95	6.63
6. Idukky	- -	- -	- -	- -	- -	- -	1	0.19	1	0.07
7. Ernakulam	30	16.67	48	15.63	85	20.19	92	17.52	255	17.79
8. Trissur	17	9.44	24	7.82	64	15.20	102	19.43	207	14.44
9. Palakkad	1	0.55	1	0.32	2	0.47	12	2.28	16	1.12
10. Malappuram	20	11.11	17	5.54	50	11.88	31	5.90	118	8.23
11. Kozhikode	1	0.55	6	1.95	15	3.56	25	4.76	47	3.28
12. Wynad	- -	- -	- -	- -	- -	- -	1	0.19	1	0.07
13. Kannur	- -	- -	- -	- -	1	0.24	7	1.33	8	0.56
14. Kasargode	3	1.67	2	0.65	8	1.90	6	1.14	19	1.32
Total	180	100	307	100	421	100	525	100	1433	100

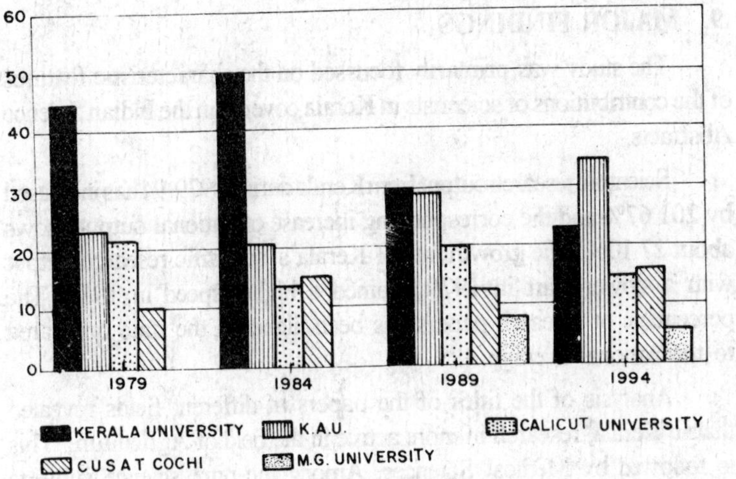

Fig. 3 Histogram showing total contribution of all Universities

In 1984, the contribution of Thiruvananthapuram increased to 64.82% followed by Ernakulam 15.63% of the total. In 1989, out of the 421 contributions of Kerala state, 164 (38.95%) were from Thiruvananthapuram, 20.19% were from Ernakulam and then came Trissur with 15.2%.

The above observation reveals that, Thiruvananthapuram was the most productive district of Kerala as far as scientific research was concerned. Its average contribution was 44.64%. Further analysis showed that it was due to the high concentration of central and state government institutions in the district.

It also showed that southern Kerala had a definite edge over Northern Kerala in the production of scientific papers.

Out of the 14 districts in the state, Pathanamthitta, Idukki, Wynad and Kannur are non-productive as far as scientific publications were concerned.

Ernakulam is the second most productive district. It is clear that, there are more central government research centres and CUSAT is also situated in the district. Trissur occupied the third position mainly due to the presence of Kerala Agricultural University.

9. MAJOR FINDINGS

The study was primarily focussed on the characteristic features of the contributions of scientists in Kerala covered in the Indian Science Abstracts.

Scientific research output from Kerala during 1979-94 has increased by 201.67% and the corresponding increase of national output shows about 27.10%. The growth rate of Kerala's scientific research output with a slower start in 1979 attained a higher speed in 1984. The percentage of research papers has been higher in the state compared to the national average.

Analysis of the titles of the papers of different fields revealed that in Kerala, research in more active in the field of Agriculture. This is followed by Medical Sciences. Among the pure science subjects Zoology comes first with largest number of contributions followed by Botany.

The study revealed that, two author papers were more in all the four years. The number of single author papers come second and the share of three author papers was below the single author papers. A noticeable trend was the constant increase in the number of papers with more than four authors.

Analysis of institutional affiliation of authors revealed that more than 50% of the papers were contributed by scientists working in Universities and Colleges.

The quantity of publications from Research Institutions was comparatively less than that of the Academic Institutions, except in the year 1984.

10. CONCLUSION

Kerala is the first State in the country to constitute a separate department for S & T and to form the State Committee on Science, Technology and Environment (STEC) for the coordination of research activities in the state. The STEC has done considerable work for the achievement of this aim, though much remains to be done.

One important area which a left out is the bibliographical control

One important area which a left out is the bibliographical control of S & T literature from the State. In order to get an idea about the scientific output from the state one has to depend on Indian Science Abstracts. But ISA can never be comprehensive as far as the state is concerned.

At present there is no means of preparing indexes or abstracts of scientific articles appearing in Malayalam. It is imperative that the STEC should take up the responsibility of the bibliographic control of the S & T publications from the state. That would go a longway in carrying out Scientometric studies which in turn would help science policy formulation in the state.

REFERENCES:

1. Abdulsalam, C A : "Where does Kerala stand among Indian states and in the world?" : a statistical study based on the composite indices of socio-economic development. - in International Conference of Kerala Studies, Abstract, Vol. 5. Thiruvananthapuram : AKG Centre for Research and Studies, 1994. p. 5.
2. Al-Kharafi, F, et al. : Science research in Kuwait: a bibliometric study.-In journal of Information Science, 13 (!), 1987. pp. 37-44.
3. Arunachalam, S : Why is Indian Science medicore?.- In Science Today, February 1990. pp. 8-9.
4. Ashokjain and Garg, K C : LASER research in India : scientific study and model projection. - In Scientomentrics, 23 (3), 1992. pp- 395-415.
5. Balachandra Rao : Kerala's contribution to Mathematics and Astronomy. - In International Congress on Kerala Studies, Abstracts, vol. 3. - Thiruvananthapuram : AKG Centre for research and studies, 1994. p. 74.
6. Braun, T, et al. : The bibliometric assessment of UK scientific performance.- In Scientometrics, Vol. 20, 1991. pp. 359-362.
7. CSIR Regional Research Laboratory : Annual Report - 1994. Thiruvananthapuram : RRL (CSIR), 1995.
8. Current trends in information technology : its impact on information science in India/edited by Kuldip Chand. - New Delhi: Batra, 1993.
9. Damodaran, V K : Development of S & T infrastructure in Kerala : a database construction report. - Thiruvananthapuram : STEC, 1991 (unpublished)
10. Denis Grogan : Science and technology : an introduction to the literature. - 4th ed. - London : Clive Bingly, 1982.
11. Egghe, L : Methodological aspects of bibliometrics. - In Library Science with a Slant to Documentation and Information Science, Vol. 25, 1988. pp. 179-91.

12. Indian Science Abstracts, 30 (1), January, 1994 New Delhi : Indian National Scientific Documentation Centre, 1994.

13. Kurian, C T : Kerala's development experience : random comments about the past and some considerations for the future. - In International Congress of Kerala Studies, Abstracts, Vol. 1. Thiruvananthapuram : AKG Centre for research and studies, 1994. p. 21.

14. Lancaster, F M : Bibliometric methods in assessing productivity and impact of research. - Bangalore : Sarada Ranganathan Endowment for Library Science, 1991.

15. Maheswarappa, B S and Ningoji, M M : Growth of literature in the field of S & T in India. - In International Information, Communication and Education (INICAE), 11(2), 1992. P. 136.

16. Measuring characteristics of scientific research : a comparison of bibliographic and survey data/by H H Garrison and others. - In Scientometrics, 24 (2), 1992. pp. 359-370.

17. Outturn of scientific and technical manpcwer in India, vol. 1 New Delhi : HRD Group, CSIR, 1993.

18. Pawan Sikka : Performance of scientific manpower in India. - In University News, 9 May 1994.

19. Pouris : A scientometric assessment of Agricultural research in South Africa. - In Scientometrics, 17 (5-6), 1989. pp. 401-413.

20. *Ranga Rao*, B V : Scientific research in India : Analysis of publications. - In journal of Scientic and Industrial Research, vol. 26, 1969. pp. 166-76.

21. Scientists and Technologists in Kerala : a Directory Thiruvananthapuram : STEC, Govt. of Kerala, 1989.

22. Sen, B K and Shailendra, K : Evaluation of recent scientific research output by a bibliometric method. - In Scientometrics, 23 (1), 1992., pp. 21-30.

23. STEC Review 1993-94 (Annual Report) Thiruvananthapuram : STEC, Govt. of Kerala, 1994

24. Turnover of science manpower declining. - In Indian Express Daily, December 3, 1993. p. 13.

25. Vasukuttan, K A : Modern science in Malayalam : a historical descriptive and critical study on technical terminology. - PhD Thesis. - Thiruvananthapuram : Dept. of Malayalam, University of Kerala, 1991 (unpublished)

26. Vijayakumar, K P : The legacy of Kerala in scientific research. - In Souvenir & Abstracts of National symposium on Environmental dimensions of Palynological Science, April, 1995. - Thiruvananthapuram : ERRC & PSI.

27. Vilanilam, J V : Science communication and development. - New Delhi : Malhotra publishing house, 1993. p-136.

8

TREND IN RESEARCH PUBLICATIONS BY INDIAN PHYSICISTS: A BIBLIOMETRIC ANALYSIS

DR. S L SANGAM &
I N NARGUND

0. INTRODUCTION

After the independence of India much importance has been given to education and research in the development of the country and a large number of Universities, Research and Development Institutions and I.I.T's were established. These institutions are engaged in the pursuit of excellence in research and education in several fields of science and technology. The literature of any field emerging form the research publications is the best indicator or exposer of the trends in the field. In the present indicator or exposer of the trends in the field. In the present paper an attempt has been made to study the trends in the field. In this paper an attempt has been made to study the trends in physics research publications and the Institutionwise contribution to the Physics research publications by analysing the research publications reported in "Indian Science Abstract, 1993".

1. OBJECTIVES OF THE STUDY

The specific objectives of the study are to :

1. identify the forms of documents in which Indian Physicists have communicated their research work.
2. study the extent of collaboration in Indian Physics literature:
3. know the institutionwise contribution to the physics literature
4. identify the scatter of Indian Physics research papers in foreign journals;
5. rank highly productive Indian journals in the field of physics.

2. RESULTS AND DISCUSSION

2.1 Forms of research communicated documents.

From the table-1 it is evident that out of the total 3167 articles published by the Indian Physicists during the year 1993, 96% of them were published in periodicals. On the other hand non serials were accounted for only 3.09% of total articles.

Table-1 Forms of research communicated documents

Forms of documents	No.of articles	%	Cumulative No.	Cumulative %
Scientific Serials	3069	96.90	3069	96.90
Ph.D Theses	91	2.87	3160	99.77
Standards	4	0.13	3164	99.90
Patents	3	0.01	3167	99.91
Total	3167	99.91		

2.2 Collaborative Trend in research publications

In recent period, there is a trend towards collaboration in research in almost all pure and applied sciences. It differs form one subject to another. From the table-2 it is evident that, in the field of physics 80% of the publications are multi authored and remaining 20% of them are single authored papers. 8748 Indian Physicists published 3069 research papers during the year 1993. The extent of collaboration in research can be measured with the help of authorship pattern of papers. According to Subramanyam[1] it can be find out by calculating the ratio of number of multi-authored papers to the total number of papers published during that year. He has given the following formula to calculate it

C = Nm/Nm+Ns

Where C = extent of collaboration;

Nm = Number of multi-authored papers; and

Ns = Number of single authored papers.

Thus the extent of collaboration is 0.80 and average authorship is 2.85.

Table-2 Collaborative Research Trends

No. of authors	No. of papers	% of papers
1	589	19.19
2	1089	35.48
3	652	21.24
4	403	13.13
5	168	5.47
6	70	2.28
7	40	1.30
8	30	0.97
9	7	0.22
10	-	-
11	1	0.03
12	1	0.03
13	1	0.03
More than 13	18	0.58
	3069	99.95

2.3 Institutionwise contribution

The table-3 shows the institutions wise contribution to physics literature. Institutionwise analysis has been made on the basis of the application of the first mentioned author. Since the non serial publications are very less in number (Table-1), therefore an attempt has been made to analyse only serial publications.

Table 3 Institutionwise contribution to physics literature

Sl. No.	Name of Institution	No. of Articles	%	Cumulation No.	Cumulation %
1	Indian Universities	1187	38.67	1187	38.67
2	Indian Research Institutions	1016	33.10	2203	71.77
3	I I T's	316	10.30	2519	52.07
4	Colleges	289	8.67	2788	90.83
5	I I Sc	162	5.27	2950	96.10
6	Foreign Universities	72	2.34	3022	98.44
7	Foreign Research Institutions	35	1.14	3057	99.58
8	Un-identified	12	0.40	3069	99.98

It can be observed form the table that for the year 1993, out of the total 3069 articles published. Indian Universities secured the first position by accounting for 38.67% of the papers. The output form the Indian research institutions ranks second with 33.10% of papers. IITs (10.30%), colleges (8.76%) and IISc (5.27%) have occupied the third, fourth and fifth rank respectively. The analysis clearly indicates that the output form the universities and research institutions together constitute more than 70% of the total output. This shows that the Indian Universities and Research Institutions contribute more to the physics literature.

2.4 Contributions by Indian Universities

Table-4 shows the extent of scatter of physics papers among different Indian Universities. In order to avoid unwieldiness, the table lists only those universities (39), contributing atleast 10 papers each. Out of 127 traditional universities in India, 91 universities have contributed atleast single paper each during the year 1993. Out of 91 Indian Universities publishing contributions in physics literature in Indian and foreign journals, 39 Indian Universities contributed 10 or more papers in the subject during the year 1993. Of these 'University of Hyderabad', had occupied the first rank by publishing 6.82% of total papers, while 'Banaras Hindu University', Varanasi has occupied second rank by yielding 5.64% of papers. 'Jadavpur University', Calcutta has witnessed 5.30% of literature, while 'Punjab University', Chandigarh and 'S. V. University', Tirupathi have contributed same amount of literature (4.21%). The other universities like 'University of Poona'; 'Devi Ahilya Vishwavidyalaya', Indore; 'University of Delhi' etc., have contributed less than 4% of literature. It is also clear form the table that, out of the total 1187 papers published by the Indian Universities during the year 1993, approximately 50% of them were shared by first 12 university, while remaining 50% by 78 universities. Thus, scattering is evident among Indian Universities and also shows the publishing habit of the physicists of most of the Indian Universities.

Table-4 Contributions by Indian Universities

Sl. No.	Name of University	No. of Articles	%	Cumulation No.	Cumulation %
1	University of Hyderabad	81	6.82	80	6.82
2	BHU, Varanasi	67	5.64	148	12.46
3	Jadavpur University, Chandigarh	63	5.30	211	17.76
4	Punjab University, Chandigarh	50	4.21	261	21.97
5	S V University, Tirupati	50	4.21	311	26.18
6	University of Poona	42	3.53	353	29.71
7	Devi Ahilya Vishwavidyalaya, Indore	41	3.45	394	33.16
8	University of Delhi	39	3.28	433	38.44
9	Cochin University of Sciences & Technology	33	2.78	466	39.22
10	Calcutta University	32	2.78	498	41.92
11	Anna University, Madras	30	2.52	528	44.44
12	Uni. of Rajastan, Jaipur	30	2.52	558	46.96
13	Shivaji Uni., Kolhapur	27	2.27	585	49.23
14	University of Madras	25	2.10	610	51.33
15	University of Mysore	22	1.85	632	53.18
16	University of Rorkee	22	1.85	654	55.03
17	Karnatak Uni., Dharwad	21	1.76	675	57.79
18	University of Allahabad	21	1.76	696	58.55
19	Andhra Uni., Visakpattanam	18	1.51	714	60.06
20	Barkatullah Uni., Bhopal	18	1.51	732	61.57
21	Himachal Pradesh University	18	1.51	750	63.08
22	Berhampur University	17	1.43	767	64.51
23	Bhagalpur University	17	1.43	784	65.94

24	JNU, New Delhi	17	1.43	601	67.37
25	Mudurai Kamraj University	17	1.43	818	68.80
26	Pondichery University	17	1.43	835	70.23
27	North Bengal Uni., Siligari	16	1.34	851	71.57
28	Osmania University, Hyderabad	16	1.34	867	72,91
29	Maharshi Dayanand Uni., Rohtak	15	1.26	882	74.17
30	University of Gorakhpur	15	1.26	897	75.43
31	Aligarh Muslim University	14	1.17	911	76.60
32	Vikram University	14	1.17	926	77.77
33	Vishva-Bharati University	13	1.09	938	78.86
34	University of Bombay	12	1.01	950	79.87
35	Manipur University	11	0.92	961	80.79
36	Bharathiar Uni., Coimbatore	10	0.84	971	81.06
37	Bhopal University	10	0.84	981	82.47
38	Sardar Patel University	10	0.84	991	83.31
39	S K University, Anantpur	10	0.84	1001	84.15
40	Other 52 Universities who have contributed less than 10 articles	186	15.66	1187	99.81

2.5 Contributions by Indian Research Institutes

Table-5 shows the scatter of physics papers among the different Indian Research Institutions. To avoid the bulkiness of the table only 19 research institutions/laboratories contributing atleast 10 papers each re enlisted. Babha Atomic Reseacrh Centre, Bombay had occupied the first rank by accounting 17.81% papers, while Tata Institute of Fundamental Research, Bombay has occupied the second rank accounting 14.07 papers. The Saha Institute of Nuclear Physics, Calcutta and National Physical Laboratory, New Delhi have published approximately the same number and occupied third rank. Thus it shows that two third of them were contributed by the first 8 institutions and the remaining one third of them were scattered among 113 research institutions and shows the poor research output.

Table-5 Scatter Physics papers among Indian Research Institute

SI. No.	Research Institution/ Laboratory	No. of Articles	%	Cumulation No.	%
1	BARC, Bombay	181	17.81	181	17.81
2	TIFR, Bombay	143	14.07	324	31.88
3	SINP, Calcutta	81	7.97	405	39.85
4	NPL, New Delhi	80	7.87	485	47.72
5	Indian Association for the cultivation of Science, Jadavpur	65	6.40	550	54.12
6	Institute of Physics, Bhubaneswar	55	5.41	605	59.53
7	IGCAR, Kalpakkam	46	4.52	651	64.05
8	PRL, Ahmedabad	24	2.36	675	66.41
9	Institute for Plasma research Gandhinagar	21	2.06	696	68.47
10	VSSC, Trivandrum	21	2.06	717	70.53
11	NCL, Poona	20	1.96	737	72.50
12	CAT, Indore	17	1.67	754	74.16
13	Institute of Mathematical Sciences, Madras	15	1.47	769	75.83
14	SN Bose National Centre for Basic Science, Calcutta	13	1.27	782	76.90
15	Bose Institute, Calcutta	12	1.18	794	78.08
16	Variable Energy Cyclotron Centre, Calcutta	12	1.18	806	79.26
17	Indian School of Mines, Dhanabad	11	1.08	817	80.34
18	NAL, Bangalore	11	1.08	828	81.42
19	Solid state Physics Laboratory Delhi	10	0.98	838	82.40
20	Other 102 Research Institution/ Laboratories who contributed 9 or less than 9 papers	178	17.51	1016	99.91

2.6 Choice of journals for publication by Indian Physicists

The considerations of a scientist in the choice of journals as reported by Ramasesan[3] suggest that a scientist chooses: 1) a journal which he feels would give him a fair criticism by competent referees; ii) a journal which deals with his own subject, preferably a theme journal, so that fellow scientists in his field can read his papers; iii) a journal where most papers are of high quality, it is anathema for a scientist to bury his paper in journals where so much is spacious or slipshot, that so valuable publications could hardly have a chance of being recognised; iv) a journal that is well edited and well printed; and vii) a journal that gives free repaints.

Rank list of scientific journals in physics was prepared on the basis of data collected form Indian Science Abstracts, 1993 is given in the table-6. Since the non serial publications were accounted for only 3.09% of titles of serial publications to prepare a core list of scientific serials in the field of physics. To prepare the rank list, the titles of journals were arranged in an order of decreasing productivity.

Table-6 Choice of journals for publication by Indian Physicists

Rank No.	Name of Scientific Journals	Country	No. of articles	%	Cumulation No.	Cumulation %
1	Proceedings of Solid State Physics Sympo-sium, BHU	India	446	14.50	445	14.50
2	Proceedings of Solid state Physics Sympo-suim BARC	India	320	10.42	765	24.92
3	Indian Journal of pure and applied Physics	India	170	5.53	935	30.45
4	Indian Journal of Physics	India	142	4.62	1077	35.07
5	Pramana	India	105	3.42	1182	38.50
6	Proceedings of Indian Science Congress, Baroda	India	87	2.83	1269	41.32
7	Physical Review B	USA	74	2.41	1343	43.73
8	Solid State Commu-	USA	70	2.28	1413	46.01

	nication					
9	Journal of Applied Physics	USA	44	1.43	1457	47.44
10	Journal of Physics-C	USA	35	1.14	1492	48.58
11	Physica C	Netherlands	34	1.10	1526	49.68
11	Physics Letters B	Netherlands	34	1.10	1560	50.78
12	Physics Letter A	Netheralnds	33	1.07	1593	51.82
13	Journal of Sound and Vibration	UK	30	0.97	1623	52.82
13	Physical Review D	USA	30	0.97	1653	53.80
14	Journal of Physics D	UK	29	0.94	1682	54.73
14	Physical Review A	USA	29	0.94	1711	55.67
15	Journal of Optics	India	28	0.91	1739	56.58
15	Journal of Physics	UK	28	O.91	1767	57.50
16	Physical Review C	USA	27	0.87	1794	58.36
17	Bulletin of Material Science	India	25	0.81	1819	59.17
17	Journal of Aeronautic Society of India	India	25	0.81	1844	60.00
18	Asian Journal of Physics	?	24	0.78	1868	60.76
18	Indian Journal of Theo-retical Physics	India	24	0.78	1892	61.54
18	Physica A	Netherlands	24	0.78	1916	62.32
18	Physics of Fulied B-Plasma Physics	USA	24	0.78	1940	63.10
19	Physica status solidi A - Applied Researcn	Germany	23	0.74	1963	63.84
19	Proceedings of Indian National Science Academy	India	23	0.74	1986	64.58
20	Physica Status solidi B - Basic Research	Germany	22	0.71	2008	65.30
21	Canadian Journal of Physics	Canada	21	0.68	2029	65.97
22	Indian Journal of Technology	India	20	0.65	2049	66.62
22	Journal Of Physics B-Atomic, Molecular and optical Physics	UK	20	0.65	2069	67.27

22	Journal of Physics G- Molecular and particle Physics	UK	20	0.65	2069	67.27
23	Applied Physics Letters	USA	19	0.61	2108	68.53
24	Journal of Plasma Physics	UK	18	0.58	2126	69.11
24	Optics Communications	Netherlands	18	0.58	2144	69.70
25	Journal of mathemetical Physical Science	India	17	0.552161 70.24		
25	Journal of Physical society of Japan	Japan	17	0.55	2178	70.80
25	Journal of the optical society of America B : Optical Physics	USA	17	0.55	2195	71.34
26	Indian Journal of applied and pure mathematics	India	15	0.48	2210	71.82
26	Metal Matter Process	?	15	0.48	2225	72.30
26	Super conductor Science and Technology	UK	15	0.48	2240	72.78
26	Zeitschrift fuer Physik C: particles and fields	Germany	15	0.48	2255	73.26
27	Journal of Instin electron Telecummun Engineers	India	14	0.45	2269	73.71
27	Zeitschrift fuer Naturforschunng Section A-Physical Science	Germany	14	0.45	2283	74.16
28	Applied Optics	USA	13	0.42	2296	74.58
28	Infrared Physics	USA	13	0.42	2309	75.00
28	International Journal of Engg Science	USA	13	0.42	2322	75.42
28	International Journal of Theoretical Physics	USA	13	0.42	2335	75.84
28	Physica B	Netherlands	13	0.42	2348	76.26
29	IETE Tech Rev	India	12	0.40	2360	76.66
29	Journal of Modern optics	UK	12	0.40	2372	77.06

29	Philosophical Matter Structural Electronic Optical and magnetic properties	UK	12	0.40	2384	77.48
29	Physics Education	UK	12	0.40	2396	77.86
29	Proceedings of Indian Academy of Science-Chemical Science	India	12	0.40	2408	78.26
30	Applied Super Conductivity	?	11	0.35	2419	78.61
30	Defense Science Journal	India	11	0.35	2430	78.96
30	Journal of Vaccum Science and Tech - Vaccum Surfaces and Films	USA	11	0.35	2441	79.31
31	Acta Mechanica	Austria	10	0.32	2451	79.63
31	Applied Radiation and Isotopea- International Journal of Radiation Applications and Instrumentation Part A	USA	10	0.32	2461	79.95
31	Bull of Calcutta Math Soc	India	10	0.32	2471	80.27
31	Bull of Indian Vaccum Society	India	10	0.32	2481	80.60
31	Nuclear Physics A	Netherlands	10	0.32	2491	80.91
31	Physics News	India	10	0.32	2501	81.23
	Journals produced less than 10 articles 192		1510	18.77	3069	100.00

Out of the total 256 scientific journals, slightly less than one half (49.68%) of total articles were covered by the first II journals which include three conference proceedings, viz. Proceedings of solid state physics symposium, BHU (14.50%); proceedings of solid state physics symposium, BARC (10.42%); and proceedings of Indian science congress, Baroda (2.83%). It is quite significant to note that approximately one fourth of articles (24.92%) were published in first two journals which were supposed to be conference proceedings of BHU and BARC respectively.

From the rank list it is clear that, proceedings of the solid state physics symposium, BHU (India) has occupied the second rank by accounting 10.42% of articles. The Indian journal of pure and applied physics has occupied third rank by receiving 6.44% of articles, while Indian Journal of Physics which also a general physics journal published from India has occupied the fourth rank by publishing 4.62% of articles, Pramana published form India has occupied the fifth rank by witnessing 3.42% of articles. Proceedings of Indian science congress, Baroda occupied sixth rank by yielding 2.83% of articles.

Physical Review-B a specialised journal in the area of condensed matter physics published from USA has occupied the 7th rank by yielding 2.41% of articles, while solid state communication has accounted 2.28% of articles and is placed in 8th rank. Journal of applied physics a pride publication of American Institute of Physics has occupied the 9th rank by accounting 1.43% of Indian physics literature, while Journal of Physics C a confirmed journal in the area of condensed matter physics published form UK has occupied 10th rank by publishing 1.41'% articles.

2.7 Extent of scatter of contribution of Indian Physicists in foreign journals

It has been mentioned in an earlier study made by Rangarajan and Gupta[2] that 'there are too many physics journals in our country and yet more than 60% of research papers from Indian physicists get published in foreign journals'. This shows the weakness of Indian physics journals. The main reasons for this trend are:

i) When any significant observation is made by any scientist, he rushes to publish his first report in a foreign journal with a view to claiming recognition among the international scientific community;

ii) Coverage of foreign journals in leading abstracting and indexing services of world reputation;

iii) Up-to datedness of foreign journals; and

iv) To highlight themselves among the scientific community of their own profession.

Table 7 Country wise distribution of Indian Physics papers

Name of the Country	Journals No.	Journals %	Articles No.	Articles %
Australia	1	0.40	7	0.22
Austria	1	0.40	10	0.32
Canada	2	0.78	22	0.71
Czechoslovakia	1	0.40	9	0.30
France	7	2.73	18	0.58
Germany	19	7.42	133	4.33
Hungary	1	0.40	1	0.03
India	52	20.31	1606	52.32
Italy	1	0.40	7	0.22
Japan	4	1.56	26	0.84
Netherlands	29	11.32	229	7.46
Norway	1	0.40	1	0.03
Poland	1	0.40	8	0.26
Singapore	1	0.40	6	0.20
Sweden	1	0.40	3	0.10
Switzerland	6	2.43	16	0.52
UK	30	11.71	283	9.22
USA	79	30.85	601	19.58
Un-identified	19	7.42	83	2.70
Total	256	100.00	3069	99.94

From the table-7 it is clear that out of the total 3069 articles published by the Indian physicists during the year 1993, 52.32% of them were published in 52 Indian Journals. While remaining 47.68% of articles were scattered in 204 foreign journals. Though Indian journals have reported slightly more number of articles, the number of foreign journals opted by Indian physicists is remarkable. So it can be inferred that, substantial amount of Indian physics literature is reported in foreign journals.

Among the journals of 17 countries in which Indian physicists have contributed, USA stands first by reporting 19.58% of articles in 79 journals, while UK stood second by reporting 9.22% of articles in 30 journals. The other major countries are: Netherlands reported 7.46% of articles in 29 journals; Germany was responsible for 4.33% of articles in 19 journals; and other countries like France, Switzerland, Japan, Canada were responsible for less than 1% of articles. The drift of Indian physicists towards foreign journals has thus made essential to take necessary remedial action to strengthen the Indian journals.

3 CONCLUSION

From the point of view of the library and information centers, it is essential to evaluate and study the research trends from time to time, so that is would be quite easy for designing, organising and managing the various information services and products to cater the information needs of researchers effectively, expeditiously and exhaustively.

REFERENCES:

1. Subramanyan, K. : Bibliometric studies as research collaboraton: a review. Journal of information science, 6: 1993. p. 35.
2. Rangarajan, K S and Gupta, B M. : Analysis of journals for publications by Indian Physicists. Journal of Library and Information Science, 4(2) 1979, pp 144-161.
3. Ramasesan, S : Proceedings of the seminar on primary communication is science and technology in India, 4-8 Dec. 1978.

9

LITERATURE ON RUBBER RESEARCH : A BIBLIOMETRIC STUDY

ACCAMMA C KORAH AND MERCY JOSE

0. INTRODUCTION

Natural Rubber, undoubtedly, is one of the most versatile vegetable products of the world. It was one of the first materials to attract the interest of the early European explorers. Hevea brasiliensis or the Para rubber tree, the principal source of this wonder product, is a native of the Amazon valley in South America, but later it found its second homes in Asian countries. Natural Rubber has proved to be a very vital industrial raw material required for the manufacture of over 35,000 different products. Thus, today natural rubber finds application in a variety of fields such as transportation, communication, agriculture, industry, health and family planning, sports and games and an umpteen number of other fields. In fact, there is not a single phase of modern life where rubber is not being used in one form or the other. Its high elastic properties coupled with excellent resilience and abrasion resistance make it the ideal choice as the most wanted industrial raw material of the day.

The cultivation of rubber is the chief means of livelihood in many countries of the Far East. The major natural rubber producing countries in the world are Malaysia, Indonesia, India, Sri Lanka, China and Thailand. The tree is also grown in Viet Nam, Cambodia, Myanmar, Bangladesh, Philippines, Papua New Guinea, Brazil, Mexico, Zaire, Liberia, Cameroon, Central Africa, Ivory Coast and Ghana. Millions of people in these countries are dependent directly or indirectly on wages or profits derived from the rubber plantations. Developed nations have explored more and more uses of rubber. In 1983, more than 7.5 million hectares of land in the world was under rubber were produced. The South East Asian region accounted for about 80% of the total production.

Among the plantation crops in India, Rubber has outpaced all others in respect of production, productivity, coverage of area and employment generation, despite the fact that it found commercial introduction to this country only as late as 1902. The accelerated pace of modernisation achieved by the rubber plantation industry is attributed largely to the intensive research and development support enjoyed by this crop. Institutionalised promotion of rubber cultivation, through dissemination of appropriate technology and extension of adequate financial and input assistance have also contributed significantly to the wide spread adoption of this crop.

Looking into the value of information on rubber and rubber products, the authors have conducted a through search on all aspects of rubber literature. Information sources such as journals and proceedings of symposia have been taken as the baseline data for the bibliometric study.

1. BIBLIOMETRICS

Bibliometrics has established itself as a viable and distinctive research technique for studying science of science based on bibliographical and citation data. Fairthorne[1] defines bibliometrics as the quantitative treatment of the properties of recorded discourse and behaviour pertaining to it. According to Nicholas and Ritchie[2], it is the statistical or quantitative description of literature. Potter[3] defined bibliometrics as "the study and measurement of the publication patterns of all forms of written communication and their authorship". Thus bibliometrics is the branch of information theory which analyses quantitatively the recorded knowledge to know its properties and behaviour. The recorded knowledge to know its properties and behaviour. The recorded knowledge may be in the form of books, periodicals, conference proceedings etc. The bibliometric studies can be divided into descriptive and behaviour studies. The present study deals with the descriptive aspects.

2. OBJECTIVES

The study was taken up with the objective of analysing authorship pattern, growth of literature and distribution of publication in different periodicals. Attempt was also made to identify core journals and to quantify growth and diversification in various facets of research.

3. METHODOLOGY

The search on rubber literature covered the period 1984-1988. Wherever possible primary journals were referred mainly; but in their absence the use of secondary journals was recorded to. Subject, author(s), year etc. with regard to each published paper were recorded on standard cards. The cards were arranged in different sequences to suit appropriate analyses.

4. RESULTS

4.1 Authorship pattern

A score of 772 communications published during the period of study was analysed to examine the authorship trend and table 1 shows the distribution of publications according to number of authors. Two and more authored papers claimed maximum score.

Table-1 Distribution of publications according to number of authors

No. of Authors	Number	Percentage
Single	316	40.93
Two and more	456	59.07

The analysis of authorship pattern indicates that single author papers account only for 40.93% and 59.07% of the papers have two or more authors. This clearly reflects the trend of team research in rubber.

4.2 Distribution of publications in various periodicals

It was seen that most of the publications on rubber were scattered in 37 periodicals. However, 19 periodicals were found to accomodate 53.5% of the papers published. The periodical 'Journal of Natural Rubber Research' (Formerly Journal of Rubber Research Institute of Malaysia) ranked first and hence was the most productive journal in rubber research during the period studied. The journals such as Rubber Chemistry and Technology, Rubber Board Bulletin, Chinese Journal of Tropical Crops etc. had got second, third and fourth positions respectively. These periodicals altogether contributed half of the research output in the discipline. Therefore these periodicals were found to be the most important journals for rubber research. The Indian Journal of Natural Rubber Research, which started publication only in 1988, was found to have the fifth position. The fact that this journal had only two issues brought out during the period of study indicates that the journal is well accepted in the field of rubber research. A list of journals selected for the study arranged in the order of their rank is given. (Table 2).

Table-2 Ranked list of journals - Journals cited more than 3 times in research papers.

Name of Journal	No. of citations	Rank No.
Journal of Natural Rubber Research	88	I
Rubber Chemistry and technology	34	II
Rubber Board Bulletin	31	III
Chinese Journal of Tropical Crops	28	IV
Indian Journal of Natural Rubber Research	25	V
Planter	23	VI
Rubber World	20	VII
Elastomerics	19	VIII
Plastics and Rubber Processing and Applications	19	VIII
Revue General Du Caoutchouc et des Plastiques	18	IX
Menara Perkebunan	17	X
Journal of RRISL	15	XI
Polymer	13	XII
NR Technology	13	XII
Bulletin Perkaretan	11	XIII
Polymer Testing	11	XIII
Pesquisa Agropecuaria Brasiliera	11	XIII
Revista Theobroma	10	XIV
Plastics and Rubber International	10	XIV
European Journal of Forest Pathology	9	XV
Journal of Applied Polymer Science	8	XVI
Planters Bulletin	8	XVI
Phytochemistry	7	XVII
Rubber News	7	XVII

Journal of Polymeric materials	6	XVIII
Agroforestry Systems	5	XIX
Rubber Chem review	4	XX
Rubber Reporter	4	XX
Progress of Rubber and Plastics Technology	4	XX
Tropical Agriculture	4	XX
Pesticides	3	XXI
Zeitschrift fur Pflanphysiologische	3	XXI
Planters Chronicle	3	XXI
Biotrop Special Publication	3	XXI
Botanical Journal of the Linnean Society	3	XXI
International Biodeterioration	3	XXI

4.3 Subject-wise Break up

Table 3 gives a subject-wise break up in the field of rubber research. The data shows that the highest percentage of papers (31.48%) fall under chemistry and technology. The second and third positions go to physiology (16.20%) and botany (15.54%) respectively. Studies on biochemical studies on rubber are the least in number.

Table-3 Subject-wise break up

Sl. No.	Subject	No. of Papers	Percentage
1	Chemistry and Technology	243	31.48
2	Physiology	125	16.20
3	Botany	120	15.54
4	Pathology	95	12.30
5	Agronomy	65	8.42
6	Economics	55	7.12
7	Biochemistry	20	2.59
8	General	49	6.35

4.4 Country-wise break up

Country-wise distribution of papers are given in Table 4.

Table-4 Country-wise break up

Sl. No.	Country	No. of Papers	Percentage
1	Malaysia	153	19.94
2	India	107	13.86
3	U.S.A.	98	12.69
4	Indonesia	60	7.72
5	France	60	7.72
6	Brazil	55	7.12
7	U.K.	45	5.83
8	Thailand	40	5.18
9	China	35	4.53
10	Sri Lanka	22	2.84

* Germany, Netherlands, Bangladesh, Ivory Coast, Mauritius, Isreal, Bangkok and Japan.

The data shows that out of the total of 772 papers, 153 are contributed by authors from Malaysia and the next place goes to India and the third to U.S.A. with 107 and 98 papers respectively. The reason for having higher production of papers from Malaysia may be the higher number of scientists working in Rubber Research Institute of Malaysia (RRIM). RRIM is the largest organization of its kind in the world devoted to research and development concerning natural rubber.

Table-5 Year-wise distribution of publications

Year	No. of Papers
1984	207
1985	195
1986	149
1987	125
1988	96

Chronological distribution of publications as shown in the above table indicates a clearcut decreasing tendency from 1984 to 1988.

5. CONCLUSION

The major findings from the above analysis are :-

1. Higher percentage of the publications are with multiple authors, indicating the trend for teamwork in research which is ideal for scientific research.

2. Most of the publications appeared in a few journals, namely Journal of Natural Rubber Research, Rubber Chemistry and Technology, Rubber Board Bulletin, Chinese Journal of Tropical Crops, etc.

3. The amount of literature on rubber pertaining to Malaysia, Indonesia, India, Thailand etc. is roughly proportional to the quantum of rubber produced in these countries.

4. 31.48% of the publications are in the field of Chemistry and Technology.

5. 19.94% of the papers are from Malaysia and the second position in the publications of papers goes to India.

6. A clearcut decreasing tendency is shown in the chronological distribution of publications during the period of study.

REFERENCES:

1. Fairthorne, RA : Empirical hyperbolic distribution (Bradford-Zipf-Mandelbrot) for bibliometric description and prediction, 1969.
2. Nicholas, David and Ritchie R : Literature on bibliometrics. London, Clive Bingley, 1978.
3. Potter, WG: Introduction to bibliometrics. Library Trends. 30; 5-7, 1981.
4. Baburajan, A : IEEE Transactions on computers: A bibliometric analysis. IASLIC Bulletin. 33(2/3); 53-59, 1988.
5. Sanjay Mishra : IASLIC Bulletin: A Bibliometric study. IASLIC Bulletin. 35(3); 125-128, 1990.

10

INDIAN JOURNAL OF AGRICULTURAL ECONOMICS: A BIBLIOMETRIC STUDY

S P USHADEVI

0 INTRODUCTION

This is an age of research and expedition in every field of knowledge. Consequent increase in the production of information is best reflected in the literature of every discipline, and economics is no exception. The case of literature that supports research activities is mainly constituted of periodicals. Growth in the number of specialised disciplines, increasing number of periodicals published in each of them and the escalating cost of this inevitable and umbiquitous medium of communication present constraints to the librarians in judiciously chalking out effective acquisition programmes for periodicals.

The field of agricultural economics indicates by its very name a combination of characteristics owned by an applied science and social science. Jouzier[1] consider it as a science of household and Marshal considers agriculture an important economic activity. According to him agricultural economics is a branch of the parent science of economics. But inter-disciplinary subjects emerging out of the fusion of traditional disciplines may also exhibit characteristics of their own. Agricultural Economics[2] is an applied phase of economics, in which attention is given to all aspects of problem, related to agriculture. Agricultural Economics is concerned with the efficient use of resources in agriculture, the technology of food and fibre production, the distribution of farm products and the income and welfare of farm population. Selection of materials of research, especially the journals, in such areas deserve greater care and understanding of the needs of the users.

An examination of the Ulrich's International Periodicals Directory shows that there are more than 117 periodicals published in agricultural economics and among them 10 are from India (Appendix I - Indian Journals in Agricultural Economics). Due to the escalating cost of the periodicals, the choice of a particular journal for a library should be done most judiciously. Indian Journal of Agricultural Economics is identified as one of the most popular one in many libraries serving advanced study and research needs on the subject. A choice made for a particular journal is not often based on any scientific study of the relative contents of the journal or the user needs. Therefore, it is relevant to study the importance of this journal through a closer examination of its contents. Moreover, a study of the citations made by the contributions to the journal is likely to throw more light on the characteristics of the materials used by agricultural economists in study and research. In addition to this, the acquisition programme of periodicals can further be guided by specific information on periodicals heavily used by agricultural economists.

1. INDIAN JOURNAL OF AGRICULTURAL ECONOMICS

Indian Journal of Agricultural Economics (IJAE) began as a biannual official organ of the Indian Society of Agricultural Economics. The first issue of the journal had been published in the year 1946. The journal was intended to publish contributions on problems of agricultural economics in India based on Scientific research investigation and first hand experience. This journal is academic in its outlook. It was also intended to provide a forum for scientific discussion among specialists in the varied topics related to Indian agriculture and rural life condition. In the year 1956 the journal became a quarterly publication. Till then there was only one issue other than the conference number.

2. OBJECTIVES OF THE STUDY

This study has been undertaken to analyse the following aspects and arrive at a conclusion regarding them.

2.1 Analysis of Papers

i) To make an analysis of the papers published in the volumes for the recent 10 years (1983-1992)., in order to collect information on the nature of Agricultural Economics and sub-topics within its purview.

ii) The number of contribution published during the decade under study.

iii) Institutional affiliation of contributors.

iv) Nature and extent of collaborative authorship.

v) Sex-wise distribution of contributors.

vi) Leading contributors

vii) The relationship between Institutional Affiliation and subject specialisation of contributors

viii) Geographical location of the contribution.

2.2 *Analysis of citations*

ix) Range (number) of cited documents and average number of reference per article.

x) Number and form of cited documents

xi) Most cited journals

xii) Geographical dispersion of cited journals

xiii) Subject dispersion of cited journals

xiv) Citation and documents recommended in the syllabus

3. HYPOTHESES

i) IJAE is rich in scholarly papers on the subject and that is the reason for the institution interested on advanced studies and research in Economics making a definite choice for this periodical to support the information needs of students, researchers and teachers.

ii) Major share of contributions to this journal comes from specialists in Agricultural Economics who are affiliated to institutions of teaching and research relevant to this field.

iii) Being an inter-disciplinary subject resulted out of the fusion[4] of agriculture and economics, a reasonable number of contributions are made from the parent disciplines too.

4. METHODOLOGY

Methodology applied in the present study involves the following:

i) Literature review

ii) Bibliometric analysis

4.1 *Literature Review*

Literature on bibliometric studies of periodicals is in abundance. But studies on specific journals are only very few.

4.2 *Bibliometric Analysis*

Bibliometric analysis is the application of mathematics and statistical methods to books and other records of communication[5]. In other words it is the quantitative treatment of the properties of bibliographic units[6].

5. ANALYSIS

The data collected are subjected to: Bibliometric analysis under the following heads:

5.1 *Number of papers*

Number of contributions in each volume varies. They fall under the range of 22-43. Ten volumes includes a total of 369 papers.

5.2 *Analysis of the corporate affiliation of the contributors revealed that the contributors mainly belonged to 3 categories*

Government

University

Others

Majority of the contributors ie. 225(60.37%) were affiliated to Government Institutions. Out of the rest 80(21.63%) belong to different Universities and 64 (17.34%) to other types of institutions like research institutes, banks etc.

5.3 *Subject-wise analysis of corporate affiliation of contributors*

Subject-wise analysis revealed that the contributors belonged to corporate bodies functioning in ten different areas.

The majority of contributors ie. (24.97%) belongs to Agricultural Economics. The next two belonged to Economics and Agriculture ie. 16.87% and 15.17%. This is followed by Economics and Social Science 10.29%, Agriculture and Rural development 8.13%, Agricultural Statistics 7.31%, Banking 7.08%, Agriculture and Technology 3.52%, Agricultural marketing 1.62% etc. In 8 cases (4.05%) subject affiliation of institution could not be identified from the information available in the journal.

5.4 Collaborative authorship

Collaboration is presumed to lead to higher quality research[7] since it often brings together a number of specialists to work on a single project. There are 282 principal authors for 369 source contributions. Majority of the contributions 197 (53.39%) are not by collaboration. 132 (35.77%) contributions are jointly produced by two authors each, 32 (8.67%) papers are contributed by 3 authors each and only 8 (2.16%) papers are contributed by more than three authors collectively.

5.5 Nature of Collaboration

Out of the 172 contributions that are having collaborative authorship, the authors of 47 (27.32%) contribution are having affiliation to institutions with the same subject field ie. Agricultural Economics. Out of the remaining 125 cases 27 (15.69%) belonged to the Economics Institutions, 25 (14.53%) belonged to Agriculture. The remaining authors are constituted to Economics and Social Science Institutions, Agriculture and Technology, Agriculture and Rural development, Agricultural Statistics Institutions, banking, Agricultural Marketing etc.

5.6 Leading Authors

From this analysis a core of productive authors were identified. Thirteen authors have contributed four or more papers to IJAE during the ten year span of the present study. These 13 authors, representing 3.49 percent of all source authors, contributed 17.2 percent of the papers which appeared in the journal. It is seen most of the papers of the leading contributors are collaborative in nature.

5.7 Institutional affiliation

Among thirteen leading contributors, three are affiliated to the same institution, Department of Economics and Sociology, Punjab (University). Two of them are affiliated to Indian Agricultural Research Institute, Delhi. Two others are affiliated to Gokhale Institute of Economics and Politics, Pune. Others are affiliated to different institutions.

5.8 Subject Specialisation

It is revealed that generally all the leading contributors are confined themselves to special areas of their own interest.

5.9 Sex of Contributors

There are a total of 427 authors (including co-authors). Out of them 403 (94.37%) are males and only 24 (5.62%) are female contributors.

5.10 Geographical Location of Contributors

As regards the number of contributors, an overwhelming majority of them, ie. 369 (86.42%) belonged to India. 58 (13.58%) contributors belonged to 13 different countries.

5.11 Subjects covered

All the source articles in the journal are included, within the frame of 15 classes. The information content of the journal in respect of this varying sub-fields of Agricultural Economics is indicated in this analysis by 3 sectors. Agricultural situation, Agricultural policy and water management and irrigation are the 3 in the primary sector, where the information value of the journal is most concentrated. The lowest attention is given to sector 3, Forestry, Economic geography and land ownership and tenure. The other nine subjects, technology in agriculture, farming, land development, farm management and allocative efficiency, agricultural business, agricultural finance, agricultural marketing, agricultural demand and supply and land reform and land use comes in the second sector.

From this analysis it is seen that most of the subject areas that discuss in the journal are very much useful to the students those who are opted agricultural economics for their post-graduate course.

5.12 Referencing Characteristics

Reference or a citation states a connection between two documents, one which cites and the other, which is cited[8]. The basic function of citing is to acknowledge work previously done by other. In the scientific tradition when a reputable scientist or technologist publishes an article, he should refer to articles which relate to his theme.[9]

The most striking feature of IJAE had been that it possessed a fairly high percentage of papers with references. Of the 369 contributions which appeared in the journal during the period of present study, 311 (84.28%) contains references. The rest 58 (15.71%) contributions were without references.

5.13 Range of cited documents

The number of references in the journal varied from paper to paper. It ranged from as little as 2 references to as many as 55 references per referenced paper. As many as 92 (29.58%) contributions contained less than 5 references. The range of 6-10 references are 78 (25.08%) contribution which is followed closely by 63 (20.25%) contributions which contained 11-15 references. As many as 32 (10.28%) contribution together contained 16-20 references. A total of 16 (5.14%) contribution contained 21-25 references.

5.14 Average number of references per article

Average number of references per source article was computed to be 8.86 (ie. 3643 cited, documents divided by 311 source articles with citation).

5.15 Bibliographic form of cited documents

The cited documents included journals, books, conference proceedings, thesis and dissertations, unpublished papers as the major forms, Reports, Government publication, Occasional papers and selected papers are brought under the group 'others'.

5.16 Most cited journals

Journals constituted the most frequently cited documents in IJAE. 237 different journals were cited. Apart from Agricultural Economics, the cited journals were associated with various disciplines such as Economics, Politics, Mathematical Statistics, Commerce and Business industry and Management, Science, History and Sociology, Demography etc. 26 journals which are cited for at least 9 times or more are selected as most frequently cited ones. The listed journals accounted for more than 70 percent of the total journal citation. It may also be significant to not that IJAE itself topped the list with 279 (18.64%) of the total journal citation. It was followed by Economic and Political Weekly, American Journal of Agricultural Economics, Econometrics, Agricultural Situation in India and other such journals.

5.17 Country-wise distribution of most cited journals

The most cited journals belong to four different countries. The country wise distribution of journals are topped by India and USA. They are followed by U.K. and Netherlands.

5.18 Geographical origin of cited journals

All the cited journals were analysed into 5 groups by (1) India (2) USA (3) U.K. (4) Other countries and (5) International bodies like

UN, ILO etc. The bulk of the cited journals belong to India (48.52%) and USA (21.93%). Out of the rest 8.43% belonged to U.K., 16.03% belonged to various other countries and 5.05% belonged to International bodies.

5.19 Subject dispersion of cited journals

Subject dispersion of cited journals has been defined as the extent to which authors in a particular discipline draw as sources out side the discipline. Low subject dispersion indicates that the cited journals belonged to a well-defined field that draws little from outside sources. High subject dispersion exist when the cited literature is largely composed of materials from outside and to which the literature belonged almost all the cited journals belonged to Agricultural Economics and its allied subject fields.

5.20 Age of cited journals

The cited journals were further examined to determine their recency. The time span between the publication of a document and its use has been examined. The analysis indicates that the current materials are important because the authors usually discuss topics of current utility and interest. As the time span increase most of the materials get obsolete. 36.02% of the cited journals were used by the contributors to this journal within 5 years of their publication, 23.33 percent within 20 years, 17.68 percent within 15 years, 12.43% within 20 years of their publication. Only 10.43% of the cited journals were consulted after more than 20 years of their publications.

6. SUMMARY AND CONCLUSIONS

This journal has a long history of nearly a half century. During the course of years the journal had expanded its scope. But within the period of this study (1983-92) not much variations in its contents are seen. During this decade the journal published 369 source articles. The largest contributors belong to Government Institutions. Next to that ranks the contributors from Universities, corporate bodies functioning in the area of agricultural economics constituted the largest contributing sector. Next comes economics and agriculture respectively. Tendency for collaborative authorship is comparatively less. 53.39% are papers of single authorship. Even among the collaborating authors the nature of collaboration is mostly conferred to two authors. The major contribution to the information content of this journal are agricultural economists working in Government institutions and University departments. Next largest contributor is economics. The most productive authors are P.K. Joshi and R.P. Singh with 8 and 7 papers respectively. Most of the papers of the leading contributors are collaborative in nature. It is revealed that generally all the leading contributors are confined to narrower areas of interest. There is a considerable predominance of male authors. Only 5.62% one female authors. It further reveals that major contributors ie. 86.72% are of Indian Origin. The rest of the contributors are from USA, Australia, Sri Lanka, Bangladesh, Canada, Pakistan, Netherlands, U.K., Holland, Japan, Malaysia, Phillipines and Singapore. Majority (35.49%) of the contributions belong to subject fields. Agricultural situation, Agricultural policy and water management and irrigation. A comparison of these subject fields with the syllabus of Agricultural economics for post-graduate course in important institutions reveals a good amount the correlation.

Of the 369 contributions in the journal 311 (84.28%) contain references. Cited documents include journals, books, conference proceedings, thesis and dissertation, unpublished papers and others. Majority of the cited documents include journals and books (41.06% and 36.34%, respectively). Majority of the contributions (54.66%) have references within the range of 0-10. Average number of references per source article is 8.86. The cited journals include journals from disciplines other than agricultural economics. The subject dispersion noted is of a smaller degree. Indian journal of Agricultural Economics is the most cited journal. A list of other 26 journals are also given. Out of the 26 most cited journals 10 are from India. Next position goes to USA. Others are from United Kingdom and Netherlands. Considering the total number of cited journals majority of them are from India. Then ranks USA and U.K. As far as the recency of the cited papers from journals are concerned. 36.02 percent of them are used within five years of their origin. Thus, more than half of the cited papers are older than 10 years. 10.43% of the papers are older than 20 years. The age of the remaining 30.14% of papers fall in between.

The journal is rich in its information content on agricultural economics. It gives emphasis to Indian agricultural economics information originating from the country, but publishes a smaller share of papers of foreign origin too. Subject dispersion in the content is very less and a large number of topics included in the special paper on agricultural economics in the institutions which focus interest on advanced studies and research in Agricultural Economics are being covered by the journal.

The scholarly nature of the journal has been improving over the years. The growth in the number and variety of communication included, the nature of papers and the background of contributors are all evidences to this. Majority of the contributors belong to government and University Departments concerned with research in Agricultural Economics or Economics. The cited literature indicates the uptodateness of information, communicated through the journal and its relevance to the Indian context.

A single journal cannot meet all the information requirements for study and research in institutions which focus interest on advanced studies and research in Agricultural Economics. However, the essentiality of continuing subscription to this journal to the related initiations is evident from the study.

REFERENCES:

1. Pal, B.W.: Economic Survey of Agriculture, Bombay: Kitab Mahal, 1961, p. 5.
2. Tuck, R.K.: An Introduction to the Principles of Agricultural Economics, Edinburgh: Neil and Co. Ltd., 1961 p.7.
3. Ulrich's International Periodicals Directory, Ed. 31, New Jersy: Bowker, 1992.
4. Ranganathan, S.R.: Prolegomena to library classification, Ed.3, Bombay: Asia Publishing House, 1967.
5. Pritchard, A.: Statistical bibliography or bibliometrics. Journal of Documentation, 25, 1969, p. 348.
6. Fairthorne, Robert A.: Emperical Hyperbolic Distribution for Bibliometric Description and Production. Journal of Documentation, 25, 1969, p. 319.
7. Fox, Mary, F. and Favu, Catherine A.: Independence and Corporation in research, journal of Higher Education, 55: 1984, pp. 347-59.
8. Martyn, John: Citation analysis: Journal of documentation, 31 (4): 1975.
9. Weinstock, Melvin: Citation indexes. In Encyclopaedia of Library and Information Science, edited by Allen Kent -nd Herold Lancour, Vol. 5, New York; Marcil Dekker, INc. 1970.

APPENDIX I: INDIAN JOURNALS IN AGRICULTURAL ECONOMICS

1. Agricultural Marketing. 1958.

2. Agriculture and Agro-Industries Journal, 1968.
3. Bulletin of Agricultural Prices, 1992.
4. Bulletin of Grain Technology, 1963.
5. Agricultural Situation in India, 1948.
6. Agricultural Banker, 1978.
7. Agricultural Wages in India, 1950.
8. Indian Journal of Agricultural Economics, 1940.
9. Rural India, 1938.
10. Rajasthan Agriculturist, 1982.

11

TUBER CROPS RESEARCH IN INDIA A BIBLIOMETRIC ANALYSIS

P K RAJAMMA

0. INTRODUCTION

"Agriculture is the mother and nourisher of all arts: when agriculture is well conducted, all other arts prosper; when agriculture is neglected, all other arts decay, on the land and on the sea". In modern times, for many countries, agriculture is a crucial source of income, employment and foreign exchange. Agriculture is closely linked with the production of the fundamental elements of man's food, shelter and clothings, derived from plants and animals and it comprises a wide variety of practices and specializations. "Agriculture has a dominant role in Indian economy contributing nearly half of the national income, providing employment to about 70% of the working population and accounting for a sizable share of the country's foreign exchange earnings".

There is no doubt that the efficient application of agricultural research is one of the primary means for accelerating the rate of agricultural development in the developing countries and for maintaining this rate at par with the advanced nations. The future of agricultural development thus, depends to a large extent on the research activities and on the research activities and on the diffusion of technology.

Indian Council of Agricultural Research (ICAR), since its inception in 1929, with its network of research institutes, research centres and research stations, plays a dominant role in the research activities of various crops. Tuber crops constitute one such group. These crops play a major role in world's food supply and have been recognized as the greatest source of dietary food energy for many of the developing countries.

1. RELEVANCE OF THE STUDY

Research in any basic or applied science depends heavily on the availability of existing information in the concerned field. Access to the precise and reliable information to the right person, in the right time is of utmost importance as this will help to minimize the wastage of resources due to reinvention, redevelopment and duplication in research work. Unless the information is directed to the target user, all the efforts for its generation become wasteful.

Journals have been playing a vital role as carriers of nascent thoughts and in the communication of scientific information ever since the publication of the first scientific journal in 1665. It has been observed that about 10-15% of the scientists working time is being spent for reading and hence an effort has been made to analyse the information gathering pattern of tuber crops research scientists in an indirect way, through the analysis of the papers published in the Journal of Root Crops which is the main channel of their communication, with a view of improving the documentation services. In a specialized research centre like Central Tuber Crops Research Institute, the information needs and use pattern of the scientists may vary according to the nature of work and level of specialization. For proper channelizing of the information to the scientists, prior understanding of the user requirements and information seeking pattern is essential. Various techniques in the analytical study help to judge these on the basis of the respective citations.

The analysis of the research output of any discipline will provide a stimulating effect on the subject and its research. As De Solla Price comments "the pattern of bibliographical reference indicates the nature of the scientific research front." It is obvious that bibliometric analysis of the prominent journals in any field can provide a major insight in to the concerned. Therefore an attempt is made in the present study to investigate tuber crops research in India, as revealed by an analysis of the Journal of Root Crops.

In the context of tuber crops research, the information generated on all aspects of tuber crops in India is disseminated through the journal, Journal of Root Crops, a half-yearly publication of the Indian Society of Root Crops, stationed at Central Tuber Crops Research Institute, Trivandrum. It has been observed that this journal carries most of the research publications on tuber crops published in India even though some appear in a few others.

2. OBJECTIVES

The major purpose of the present study is to examine the studies and research conducted during the last nineteen years on various aspects of tuber crops and can be summarised as follows:

* to ascertain the strength and weakness of the research activities conducted during the last nineteen years on tuber crops
* to draw a detailed picture of the way in which literature has developed during the period of study and the relative distribution in the various facets of the subject field.
* to identify the gaps in the research output in respect of various forms of tuber crops and their allied areas
* to identify in quantitative terms, the relative use of the different forms of documents, authorship pattern, language etc. and their chronological scattering.
* to understand the core journals for the scientists working on tuber crops and to identify them
* to investigate the degree of collaboration in authorship in the field of tuber crops research

3. SCOPE AND LIMITATIONS

The present study is confined to the literature published in the Journal of Root Crops, which ofcourse, carry the majority of the research output on tuber crops, for a period of nineteen years as stated above. However the findings are likely to have general implications for the scientific community working on tuber crops. The period of the study is 1975-1993 and for the citation analysis, an interval of five years is selected as the total number of references for the total nineteen years is much larger.

4. METHODOLOGY

The following methods were adopted for the present study.

i) Literature survey

ii) Citation analysis

i) Literature survey

In order to identify similar studies on agriculture, the LISA Plus 1994 on CD ROM was searched. In addition to this, some of the primary periodicals in the field of Library and Information Science were also scanned.

With the background information obtained from these, the papers published in the Journal of Root Crops for a period of nineteen years, 1975-1993 were taken for the study. Entries were made for the 450 papers published in the regular issues of the Journal of Root Crops for the above period on 5"x3" slips. The details such as author(s), title, year, total number of references appended to etc. were recorded. These slips were then put to different analysis.

ii) Citation analysis

According to Garfield "true citation analysis is one which deals with works cited as having actually been used in the preparation of, or having otherwise contributed to the source paper". In the present study, this method is used for measuring the relative use of journals as sources of research information and also for identifying the core literature of specialised area or discipline.

The 450 papers generated 4003 citations and as this was a large number to study the details, the analysis was five years ie, 1975, 1979, 1989 and 1993.

5. TUBER CROPS RESEARCH

Tuber crops represent a group of crops which accumulate carbohydrates in the enlarged underground, modified stems or roots. As per the Root Crops Digest, the term 'root crops' is also used to represent these crops. Accordingly, these constitute plants which produce subterranean structures that may be used for human or animal food. They are normally perennating organs, storing plant nutrients through a resting period (dry season or winter) which are used in the regrowth of the plant when the growing conditions are again favourable. The term "root" is often a misnormer, as in many cases the storage organ may be a morphologically a modified stem, eg. a swollen rhizome or corm, or a tuber such as potato, a swollen underground organs are commonly known as 'tubers'. The important tuber crops grown in India are the following.

Subgroup	Crop name	
	Common	Botanical
Major crops	Cassava	Manihot esculenta
	Sweet potato	Ipomoea batatas

Yams	Greater yam	Dioscorea alata
	Lesser yam	D. esculenta
	White yam	D. rotundata
Aroids	Elephant foot yam	Amorphophallus paeonifolius
	Giant taro	Alocasia indica
	Tania	Xanthosoma sagittifolium
	Taro	Colocasia esculenta
Minor tubers	Arrow root	Maranta arundinacea
	Chinese potato (Coleus)	Solenostemon rotundifolius
	Winged bean	Psophocarpus tetragonolobus
	Yambean	Pachyrrhizus erosus

The major tubers include a number of others of local importance.

The tropical tuber crops play a key role in the world's food supply. These are consumed by a third of the world's population, usually comprising the lower socio-economic groups. The consumption rate of tropical tuber crops compare very favourably with that of cereals throughout the world. These crops have been recognized as the greatest source of dietary food energy for majority of the developing countries. Apart from the importance of these crops in food security and food self-sufficiency, they contribute significantly to the national economies and rural income and provide employment to the people.

Research on tuber crops at global level is being carried out in many of the international research centres. A major role in the conservation of germplasm of these crops and other concerned activities are conducted by the Consultative Group on International Agricultural Research (CGIAR). Other institutions like Centro International de Agricultural Tropical (CIAT), Columbia, International Institute of Tropical Agriculture (IITA), Nigeria, Centre International de is Papa (CIP), Peru, Asian Vegetable Research and Development Centre (AVRDC), Taiwan etc. also are engaged in the research on the various aspects of these crops.

6. TUBER CROPS-INDIAN SCENE

The establishment of the Imperial Department of Agriculture (Calcutta) in 1881, followed by the emergence of State Department at Nagpur during the period 1883-1895 were the milestones in the beginning of scientific farming in India. The Government of India laid the foundations of the Imperial Institute of Agriculture at Delhi in 1905 and after independence, it was renamed as Indian Agricultural Research Institute. Imperial (now Indian) Council of Agricultural Research was established in 1929 as a registered society. The period 1921-1962 witnessed the setting up of various Central Commodity Committees dealing with research on specific crops which located in the regions of the crops concerned. Various research institutes were also started. Simultaneously, state governments also set up their own research stations and experimental farms for various crops.

The starting of agricultural universities is also an important landmark in the history of agricultural research in India. The first agricultural university came into existence at Pantnagar (UP) in 1960 and now there are 27 state Agricultural universities established throughout India. These universities with their colleges and research departments contribute much to the development of agriculture in the country.

It was in this context that the Travancore University at Trivandrum established a tuber crops research station for conducting and co-ordinating research on tuber crops. Realizing the importance of these crops in the economy of India, Indian Council of Agricultural Research (ICAR) financed the researches of the above station through adhoc research schemes and later in 1964, the research station was designated as Central Tuber Crops Research Institute (CTCRI).

A good research base exists at present in the country to sustain root and tuber crops development. It consists of the CTCRI and the All India Co-ordinated Research Project on Tuber Crops (AICRPTC).

Central Tuber Crops Research Institute (CTCRI), Trivandrum, Kerala.

This institute is the only one of its kind in the world exclusively dedicated to the development of tuber crops. The main aim of the establishment of the institute was to conduct and co-ordinate research on tuber crops in the country. The institute has a Regional Centre at

Bubhaneswar, established in 1976. The all India Co-ordinated Research Programme was initiated during 1967 and presently the project has ten centres spread all over the country. CTCRI is also co-operating as the centre for All India Co-ordinated Project on Harvest and Postharvest Technology.

The important crops dealt with at the Institute are Cassava (Mani hot esculenta Crantz), sweet potato (Ipo-Burk, D. rotundata Poir), aroids Taro (Colocasia esculenta (L) Schott), elephant foot yam (Amorphophallus paeonifolius (Denst) Nicolson) and minor tuber crops like coleus (Solenostemon rotundifolius), Yambean (Pachyrrhizus erosus (L). Urb) and arrow root (Maranta arundinaceae L.)

Research in the institute is conducted in various divisions viz. Division of Crop Improvement, Division of Crop Production, Division of Crop Protection, Division of Crop Utilization and Biotechnology and Section of Social Sciences. The Regional Centre located at Bubaneswar is concentrating on location specific problems mainly in sweet potato and aroids. The headquarters of the All India Co-ordinated Project is also located at the institute. The administrative, technical and farm wings of the institute carry out the related support activities for the research.

Tuber Crops literature

Tuber crops represent only a part of the root crops under the major division of field crops. The percentage of literature output on these crops in the context of the world's total agricultural literature is very meagre. This fact is clear from an analysis of the coverage of tuber crops in the major global agricultural database Field Crops Abstracts, brought out by CAB International, UK. A perusal of the figures for a period from 1975-1993 showed that the papers on tuber crops represent only about 3% of the total output and in most cases it is below 2%.

Information sources in tuber crops

It is observed that information sources in applied sciences, particularly in agriculture is greater than in pure sciences, and hence the publications appear in many forms like journals, monographs, bulletins, conference/workshop/symposia proceedings, theses, reports, abstracts, serials like annual reviews, statistical serials, grey literature, press clippings etc. In the case of tuber crops also, information is being disseminated through many kinds of documents. Apart from these, scientists use time series data sources, addresses and proceedings of research group discussions, newsletters, circulars etc.

Bibliographic control

A major problem confronted by the users of tuber crops literature is the lack of bibliographic control at the national and international level. At the international level, agencies like FAO and CAM International, UK are providing useful services in agriculture viz. Agris, CAB Abstracts etc., but the coverage of tuber crops in these databases are poor or not adequate.

Major centres engaged in tuber crops research are bringing out their publication lists and related services. In this context it is worth mentioning that Abstracts on Cassava, a comprehensive database on cassava having world coverage has been brought out by CIAT, Columbia.

Being the national centre engaged in tuber crops research, CTCRI is also contributing in this respect. In this regard, a bibliography on Tuber Crops research in India has been released. It provides 2225 citations of tuber crops research papers published by Indian scientists in India and abroad.

7. ANALYSIS

Bibliometric analysis is used to find out the nature and characteristics of tuber crops research, based on the papers published and the citations appended to these papers in Journal of Root Crops, for a period of nineteen years (1975-1993).

The analysis was done in two steps. Firstly, the 450 papers were studied in detail. Secondly, the citations appended to the papers in the journal were analysed in detail, at an interval of five years.

Analysis of the papers.

In the first level of analysis, all the details such as author(s), title, year, total number of references appended to, etc. of the 450 papers were put to various analysis.

Year-wise analysis of the papers

The number of papers and their corresponding year are shown in the table 1.

Table-1 Year-wise distribution of papers 1975-1993

Year	No. of papers	Percentage	Cumulative %
1975	16	3.56	3.56
1976	27	6.00	9.56
1977	33	7.33	16.89
1978	28	6.22	23.11
1979	22	4.89	28.00
1980	18	4.00	32.00
1981	18	4.00	36.00
1982	14	3.11	39.11
1983	13	2.89	42.00
1984	17	3.79	45.79
1985	12	2.67	48.46
1986	19	4.22	52.68
1987	22	4.89	57.57
1988	29	6.43	64.00
1989	30	6.66	70.66
1990	34	7.56	78.22
1991	34	7.56	85.78
1992	34	7.56	93.34
1993	30	6.66	100.00
Total	450	100.00	

The above table showed that the maximum number of papers were published during the period 1990-1992 (34 papers/volume each year) and the minimum appeared in the year 1985 (12 papers/volume). Though there is no spectacular growth in the number of papers in recent years, their number is definitely increasing. While it took 11 years to publish roughly half of the total number of papers (to be exact 48.46 percent), the rest of them were published in 8 years.

Crop-wise break-up of papers

The papers were sorted on the basis of the specific crops. The results are shown in table 2.

Table-2 Crop-wise bréak-up of papers 1975-1993

Crop	No. of papers	Percentage	Cumulative %
Cassava	178	39.56	39.56
Sweet Potato	111	24.67	64.23
Yams	50	11.11	75.34
Aroids	73	16.22	91.56
Minor tubers	19	4.22	95.78
Root crops general	12	2.66	98.44
Others	7	1.56	100.00
Total	450	100.00	

It is seen from the table that 39.56% of the papers were about cassava, 24.67% relate to Sweet potato, Aroids from 16.22%, Yams 11.11%, Minor tubers 4.22%, Root crops general 2.66% and other tuber crops constitute 1.56%.

The major crops dealt within the Central Tuber Crops Research Institute are Cassava and Sweet potato and this aspect is reflected in the case of publications also. As a result, the number of papers appearing on other crops are comparatively less. These crops also deserve due considerations in the context of research and publication for popularising them as they also form important sources of starchy food.

Subject-wise analysis of the papers

The papers were sorted on the basis of the major subject headings. The subject groups were crop improvement, crop production, crop protection, crop utilization and general aspects.

Crop improvement includes the papers on genetics, cytology, plant breeding and other research on the varietal improvement of tuber crops. Crop production covers papers dealing with cultivation, agronomic aspects, fertilizers, irrigation etc. and physiological aspects of crop production. Crop protection consists of the pathological, entomological and plant protection aspects. Crop utilization refers to the use of these crops for human or animal use.

Analysis of the papers on the above aspect showed the details as in table 3.

Table-3 Subject-wise break-up of papers

Subject	No. of papers	Percentage
Crop improvement	102	22.67
Crop production	218	48.44
Crop protection	83	18.44
Crop utilization	30	6.67
General	17	3.78
Total	450	100.00

The table revealed that the highest percentage of papers are in the crop production group with 218 papers (48.44%). Next is the crop improvement with 102 papers (22.67%), crop protection (18.44%), crop utilization (6.67%) etc. This may be due to the fact that the crop production aspect is given much stress in the context of research.

Authorship Pattern

Multiple authorship has been a characteristic feature of the modern science and there has been a consistent trend towards increased collaboration in all the branches of sciences. It has also been found that the rate of increase in the multiple authorship varies considerably with the subjects concerned. In order to find the nature of authorship pattern in the case of tuber crops research, the entries prepared for the papers were put to study.

The number of papers with 1, 2, 3, 4 and more than 4 authors and the total number of papers in each year for the Journal of Root Crops is given in the Table 4.

Table-4 Authorship pattern of the papers 1975-1993

Year	Number of authors					Total
	1	2	3	4	More than 4	
1975	5	5	6	-	-	16
1976	5	9	11	2	-	27
1977	6	16	9	2	-	33

1978	6	13	7	2	-	28
1979	2	16	3	1	-	22
1980	3	7	6	2	-	18
1981	4	12	2	-	-	18
1982	2	5	5	2	-	14
1983	1	8	4	-	-	13
1984	2	8	5	2	-	17
1985	1	6	4	1	-	12
1986	2	6	7	4	-	19
1987	5	8	5	2	2	22
1988	8	7	10	6	1	29
1989	6	13	8	1	2	30
1990	5	16	11	2	-	34
1991	4	17	6	4	3	34
1992	4	15	19	5	1	34
1993	1	13	7	7	2	450
Percentage	15.33	44.45	27.	10.00	2.44	100

The table showed that the maximum number of papers were with two authors (44.45%), papers with three authors (27.78%) occupied the second positions followed by single authored papers (15.33%). Papers with more than four authors formed only 2.44%. Two and Three-author papers constituted more than 72% of the total. Another finding is that papers with more than 4 authors were conspicuous by their absence until 1986.

Degree of Collaboration in research

The degree of collaboration in research was found, using the formula stipulated by Subramanyam. The formula is:

C = Nm/Nm + Ns

where C = degree of collaboration in a discipline

Nm = Number of multi-authored research papers in the discipline published during a year.

Ns=Number of single authored formula, the degree of collaboration in the field of tuber crops research has been determined for the period 1975-1993, for every alternate year and is given in the Table 5.

Table-5 Degree of Collaboration

Year	Papers		Total No.	Degree of
	S.author	M. author	of papers	collaboration
1975	5	11	16	0.68
1977	6	27	33	0.81
1979	2	20	22	0.90
1981	4	14	18	0.77
1983	1	12	13	0.92
1985	1	11	12	0.91
1987	5	17	22	0.77
1989	6	24	30	0.80
1991	4	30	34	0.88
1993	1	29	30	0.96
Total	35	195	230	

The table shows the maximum collaboration of the authors for the papers published in the year 1993 (0.95%). The analysis showed that the degree of collaboration in the case of the above journal increased from 0.68 in 1975 to 0.96 in 1993.

Tuber crops research papers show a trend of increasing multiple authorship and decreasing single authorship.

Institutional Collaboration of the Authors

A detailed study of the institutional collaboration of the authors who contributed papers to Journal of Root Crops for the period 1975-1993 was made and the results are shown in the table 6.

Table-6 Institutional Collaboration of the Authors

Sl. No.	Name of the University/Institution	No. of authors	Percentage
1.	Central Tuber Crops Research Institute, Trivandrum	654	60.44

2.	Kerala Agricultural University	125	11.55
3.	Bidan Chandra Krishi Viswa Vidyalaya, West Bengal	42	3.88
4.	Tamil Nadu Agricultural University, Coimbatore	32	2.96
5.	Bangladesh Agricultural Research Institute, Gazipur	27	2.50
6.	Obafami Awolowo University	26	2.40
7.	Rajendra Agricultural University, Muzaffarpur	24	2.22
8.	NBPGR Regional Station, Trichur	24	2.22
9.	University of IFE, Nigeria	21	1.94
10.	National Root Crops Research Institute, Nigeria	19	1.76
11.	University of Agricultural Science, Bangalore	12	1.11
12.	Konkan Krishi Vidyapeeth	10	0.92
13.	University of Ibadan, Nigeria	8	0.74
14.	Central Food Technological Research Institute, Mysore	7	0.65
15.	Marathwada Agricultural University	7	0.65
16.	Orissa University of Agriculture and Technology	7	0.65
17.	Natural Resources Institute, London	7	0.65
18.	Birsa Agricultural University, Ranchi	6	0.55
19	ICAR, North-Eastern Hill Region Research Station	6	0.55
20.	Regional Research Laboratory, Trivandrum	6	0.55
21.	International Institute of Tropical Agriculture, Nigeria	5	0.46
22.	Others	7	0.65
	Total	1082	

The table showed that out of the 1082 authors, 60.44% are scientists of CTRCI. 11.55% comprises researchers from the Kerala Agricultural University and its affiliated colleges and research stations. Other national and international organisations are also coming to the scene, as revealed in the above table.

It may be noted that though the Journal of Root Crop is the main outlet of tuber crops research in India, it is also used by other scientists in the field to publish their research work, however, the major contribution is from India, mainly from CTCRI.

Country-wise break up of authors

The country-wise analysis of the authorship showed that 88.9% of the contribution to the journal is from India. Next came Nigeria with 7.3% and Bangladesh 2.5%. United Kingdom contributed 0.65% and other countries 0.65. The details are shown in the table 7.

Table-7 Country-wise break-up of authorship

Country	No. of authors	Percentage
India	962	88.90
Nigeria	79	7.30
Bangladesh	27	2.50
UK	7	0.65
Others	7	0.65
Total	1082	100

8. ANALYSIS OF CITATIONS

Citation analysis is used in the present study to understand the nature and characteristics of tuber crops research. Citations are references appended to scientific papers, notes, reviews, correspondences etc. published in scientific journals. The analysis of these citations have been established as a method of studying the journals as well as the people and the work of science. The citation links provide a quantitative picture of the journal utility and relationships that are useful in many ways. Here, the conclusions are mainly based on what scientists say about the utility and relationships in their choice of references. These varies according to the function of the journal.

The primary purpose of citation is to enable a reader to go to the referred to document for information on a point or check the authenticity of a particular view, finding or method. Each citation is a message from the author of a document to his readers.

Citations also serve other purposes.

* as a bibliography
* for preparing ranked list of periodicals
* for understanding the relative use of different types of documents
* for calculating the useful life of documents
* to find out the relatedness and dependence of subjects
* to calculate the citation rate of journals
* to find the impact factor for a concerned journal
* to calculate the immediacy index etc.

The 450 papers published in the Journal of Root Crops generated 4003 citations. But, for the present study, the citations were collected at intervals of 5 years viz. 1975, 1979, 1984 and 1993 and thus a total of 1107 citations were collected from the 115 papers published during the above periods. All the 4003 citations could not be subjected to analysis mainly due to shortage of time.

Entries were made for the citations on 5" x 3" slips, indicating the relation between the cited and the citing document. Details of the cited author(s), cited paper, source of the cited paper, year of the cited paper, citing author, title of the citing paper, source of the citing paper including the pages, year of the citing paper etc. were recorded and thus 1107 slips were prepared and these were subjected to different analysis.

Distribution of citations and the average citation rate

The 450 papers published in the Journal of Root Crops generated 4003 citations. The number of papers and the total number of citations in each year is shown in Table 8.

Table-8 Distribution of papers and citations 1975-1993

Year	No. of Papers	No. of references	Average	Percentage
1975	16	258	16.13	6.45
1976	27	215	7.96	5.37
1977	33	286	8.67	7.15
1978	28	265	9.46	6.62
1979	22	171	7.77	4.27
1980	18	110	6.11	2.75
1981	18	147	8.17	3.67
1982	14	144	10.29	3.59
1983	13	193	14.85	4.82
1984	17	105	6.18	2.62
1985	12	140	11.67	3.50
1986	19	171	9.00	4.27
1987	22	172	7.82	4.29
1988	29	258	8.90	6.45
1989	30	218	7.27	5.45
1990	34	233	6.85	5.83
1991	34	282	8.29	7.04
1992	34	280	8.24	6.99
1993	30	355	11.83	8.87
Total	450	4003	8.89	100.00

The number of citations in volume were found to vary from 105 to 355. While the year 1993 was responsible for the highest number (8.87%), 1984 accounted for the lowest number (2.62%). Apart from this, no consistent pattern of increase or decrease in the number of citation is discernable. The table also showed that the highest average of citations were for the papers published during the period 1975, i.e. 16 papers. The year 1980 had only 110 citations registering an average of 6.11.

The average citation rate for the period 1975-1993 was found to be 8.89.

Type of documents cited

The slips were sorted into different categories on the basis of documents cited. They were grouped into journal articles, books, proceedings of workshops/symposia, reports, bulletins, serials, theses and others which do not belong to any of the above groups. These represent the major source of references though some other like newspaper articles, patents, standards, circulars, etc. were found to be used by the scientists.

The papers that appeared in research journals and popular periodicals constitute journal articles.

The group, books, denote trade books and monographs other than serials.

Proceedings of workshop/symposia cover the papers presented to the workshops and symposia as well as published proceedings of these.

Report refers to the research reports and progress reports of research organizations and institutions including annual reports.

Bulletins are those issued by specific institutions containing special information and are often oriented towards a particular group of authors.

Serials are publications at regular or irregular intervals in which each volume embodies the same kind of information, presented in the same pattern, relating to its year of coverage. It is usually issued as a complete volume, mostly by corporate bodies.

Details of the type of documents consulted by the tuber crops scientists given in the table 9.

Table-9 Type of documents cited

Type of documents	Citations	Percentage
Journal articles	692	62.51
Books	149	13.46
Proceedings	120	10.84
Reports	54	4.87

Bulletins	32	2.90
Serials	26	2.35
Theses	22	1.99
Others	12	1.08
Total	1107	100.00

The study revealed that journals are the preferred source of information (62.51%). Books were found to be of less importance to them as their percentage of citation was only 13.46. Scientists were found to consult the proceedings of workshops/symposia almost equal to books (10.84%) and bulletins constituted 2.9%. Theses were found to be less used (1.99%).

Chronological analysis of citations

The citations were sorted in the chronological order and the results are represented in the Table 10

Table-10 Chronological analysis of Citations

Period	Citations	Percentage
1901-1910	6	0.54
1911-1920	5	0.45
1921-1930	11	0.99
1931-1940	54	4.88
1941-1950	52	4.70
1951-1960	115	10.39
1961-1970	222	20.05
1971-1980	379	34.24
1981-1990	228	20.60
1991-1993	35	3.16
Total	1107	100.00

It was found that the earliest citations date as back as the period 1901-1910. There is a constant increase from these upto 1980. After this period, there is a sudden decrease. The maximum number of citations belonged to the period 1971-1980, accounting for 34.24%. It can be seen that about 78% of the citations were from the period 1961 onwards. Though the scientists use fairly recent publications, there is a time lag of a few years in citing the publications.

Journal articles

The 692 journal articles obtained were analysed in detail. The most frequently used journals by the tuber crops research scientists were identified and the ranked list of journals were prepared, based on the citation rate of the journals.

Table-11 Ranked list of cited journals

Rank No.	Journal	Citations	Percentage
1	Journal of Root Crops	63	9.10
2	Journal of the American Soc. for Hort. Science	30	4.33
3	Chemical Abstracts	20	2.89
4	Review of Plant Physiology	19	2.75
5	Experimental Agriculture	15	2.17
6	Plant Physiology	14	2.02
7	Crop Science	13	1.88
7	Tropical Agriculture	13	1.88
8	Indian Journal of Genetics and Plant Breeding	10	1.46
8	Phytopathology	10	1.46
9	East African Agriculture and Forestry Journal	9	1.30
10	Annals of Botany	8	1.16
10	Dissertation Abstracts International	8	1.16
10	Madras Agricultural Journal	8	1.16
10	Oecologia	8	1.16
11	Agric. Research Journal of Kerala	7	1.01

11	Annals of Applied Biology	7	1.01
11	Annals of Tropical Research	7	1.01
11	Bragantia	7	1.01
11	Indian Journal of Agronomy	7	1.01
11	Indian Journal of Agricultural Science	7	1.01
11	Nature	7	1.01
12	Field Crops Abstracts	6	0.87
12	Plant Disease Reporter	6	0.87
12	Tropical Pest Management	6	0.87
	Above 12th rank	377	54.48
	Total	692	100.00

The list of the Journal (table 11) showed that the Journal of Root Crops is the most highly used Journal as far as the tuber crops research is concerned. This finding support the earlier findings that scientists of highly specialised disciplines depend on the core journals in the specific discipline for their information requirements. Tuber crops research is a narrow field in agricultural sciences and journals devoted to this area are very few in number. As the Journal of Root Crops forms the main outlet for tuber crops research, it is also the main source of reference to those who are engaged in tuber crops research. This journal occupied the first position in the rank list with 63 citation, indicating a high self-citation rate. The second one, ie., the Journal of the American Society for Horticultural Science is having 30 citations and is an established journal in the field of crop sciences. The third journal in the rank is Chemical Abstracts and Review of Plant Pathology; both are secondary periodicals in the field of agriculture. A high use of the secondary sources has been indicated, especially for the period before 1950, may be due to the absence of non-availability of original papers.

Out of the 236 journals cited, it was seen that 35 have been frequently cited by the scientists. The list prepared showed that most of these are journals related to agriculture sciences in varying degrees and most of these are available in the CTCRI Library.

Bradford's distribution

A table with the cumulative number of journals and citation was prepared to verify the Bradford's Law of Distribution in the case of tuber crops research.

Table-12 Distribution of Journal citations

Cumulative No. of Journals	Cumulative No. of Citations
1	63
2	93
3	113
4	132
5	147
6	161
8	187
10	207
11	216
15	248
22	297
25	315
35	365
49	421
66	472
116	572
236	692

The data gives in the above table indicate that the citations agree with the Bradford's Law of Scattering. The law states that "articles of interest to specialist must occur not only in the periodicals specialising in his subject, but also in other periodicals, which grow in number as the relation of their fields to that of his subject lessens and the number of articles on his subject in each periodical diminishes". Bradford concluded that if periodicals are listed in the decreasing order of productivity of papers on a subject, then they stand divided into a nucleus of periodicals more devoted to the subject and several other zones when the number of periodicals in the nucleus and the succeeding zones will be in the ratio 1:N:N^2.

When the total number of 692 citations were divided into three zones, each one containing nearabout 230 numbers, the proportion of the journals in each zone was found to be 11:55:170. The corresponding proportion of citations in each zone was 216:256:220. It is obvious that when the number of journals in the first zones perfectly matches with that of Bradford (assuming that the value of N is about 5), the number is less in the third zone. Ideally the number would have been 11:55:275. This gave the indication that there are more journals in the peripheral zone which are not commonly used the scientists.

Chronological analysis of Journal citations

The citations were grouped based on the year of publication into a period of ten years as shown in the Table 13

Table-13 Chronological analysis of Journal citations

Period	No. of citations	Percentage
1901-1910	1	0.14
1911-1920	-	-
1921-1930	15	2.17
1931-1940	45	6.50
1941-1950	60	8.67
1951-1960	65	9.39
1961-1970	192	27.76
1971-1980	173	25.00
1981-1990	105	15.17
1991-1993	36	5.20
Total	692	100.00

The analysis showed that the maximum number of citations belonged to the period 1961-1970 (27.76%) and there was a trend of decrease in the number from 1971 onwards.

This may be due to the fact that the period 1961-1980 records the maximum of research conducted in all the basic areas of tuber crops and a number of publications were released during this period. It is seen that these papers are being quoted frequently by the teachers in the field even in the latest instances.

Geographical analysis of Journal citations

A country-wise study of the journal citations indicated that 28.32% of the citations were from USA, 27.46% from India and 20.52% from UK. The details of the analysis are shown in the Table 14.

Table-14 Country-wise break-up of journal citations

Country	Citations	Percentage
USA	196	28.32
India	190	27.46
UK	142	20.52
Netherlands	51	7.37
Puerto Rico	30	4.34
Canada	19	2.75
China	15	2.17
Philippines	14	2.02
South Africa	9	1.30
Japan	7	1.01
Denmark	4	0.58
Germany	4	0.58
Others	11	1.58
Total	692	100.00

Other types of documents

As the number of citation in the other types of documents were found to be very meagre, further analysis of theses were not conducted.

9. FINDINGS AND CONCLUSIONS

The present study was initiated with the main objectives of examining the research work done on various aspects of tuber crops and to reveal the information seeking pattern of the scientists working on tuber crops by way of their citations to documents in their variety of forms, age etc. As the research output generated on different facets of tuber crops research is being disseminated through the academic journal, Journal of Root Crops, the investigator presumed that an indepth analysis of the papers and respective citations would definitely reveal the characteristics of tuber crops research and the relative use of information sources. All the papers published in the journal for the period, 1975-1993 and the citations appended for these published papers for the interval period of five years ie. 1975, 1979, 1984, 1989 and 1993 were put to extensive analysis. The investigation revealed many characteristic features of the research in this field identifying the strength and weakness of the research process over the period and how the information is used by their respective scientists. This study revealed many characteristic features of the research in the field, which may be considered to represent the research scene in India. The major findings are the following.

Number of Papers

In the case of the number of papers published in the journal, the maximum has been recorded during the period 1990-1992 (22.68%). The number is definitely increasing as it took 11 years to publish approximately half of the total number of papers (48.46%), the rest of them were published in just eight years.

Major crops covered

Cassava and sweet potato contribute 39.56% and 24.67% of the papers respectively. Aroids occupy the third (16.22%) and Yams (11.11%) the fourth position. This is an indication that research in the field of tuber crops is mainly centered around cassava and sweet potato. Other crops like aroids, yams etc. are not considered equal or not given due consideration. The trend is similar in the international scene where majority of the papers published belong to cassava and sweet potato. Papers on other crops are comparatively less.

Subject - analysis

Subject-wise study of the papers showed that they deal with the following aspects.

Crop production	48.89%
Crop Improvement	23.56%

Crop protection	17.33%
Crop utilization	6.44%

The above trend is seen in all the crops except in the case of papers on yams where the area crop improvement shows priority over crop production as far as the number of papers are concerned.

Authorship Characteristics

Multiple authorship, a distinctive feature of the modern science is exhibited in the case of tuber crops research also. Maximum number of papers were two - authored (44.45%) and three-authored papers constituted 27.78%. These two categories formed more than 72% of the total publications. Tuber crops research papers further show a trend of increasing multiple authorship and decreasing single authorship.

Collaborative research

Degree of collaboration of the authors for the journal papers increased from 0.68 in 1975 to 0.96 in 1993. This is similar to the field of Agriculture, Biochemistry, Zoological Sciences etc. where studies have showed the increase of multiple authorship.

Institutional affiliation of authors

Institutional collaborative studies showed that 60.44% are scientists of CTCRI, 11.55% comprises researchers from Kerala Agricultural University and its constitutional colleges. Apart from these two institutions, researchers from national and international organizations have contributed papers to the journal (28.01%).

Geographical distribution of authors

88.9% of the total contributors came from India. Country like Nigeria, Bangladesh, UK etc. are also found. It was seen that no scientist from developed countries, except UK has contributed papers to Journal of Root Crops.

Number of Citations

The number of citations in a volume was found to vary from 105 to 355. The year 1993 was found to be for the highest number (8.87%) and 1984 accounted for the lowest number of citations (2.62%).

The average citation rate of the Journal of Root Crops for the period 1975-1993 was found to be 8.89%.

Journals - the main medium

The study revealed that journals were the preferred source of information to scientists working on tuber crops (62.5%). Books were found to be less used by them as the percentage of citations for books was only 13.46%. Apart from these, scientists were found to use proceedings of workshop/symposia, bulletins, reports, serials etc. in varying degree of preferences. This finding resembles the instances in pure sciences where the journals form the chief source of information and the medium for communicating the research findings.

Chronology of the citations

It was observed that though the scientists made use of recent information in the concerned field, there was a time lag for citing these. The maximum number of citations analysed belonged to the period 1971-80 (34.24%)

Ranking of periodicals

The ranked list of journals prepared from the journal citations revealed that the Journal of Root Crops was the most highly used journal in the field. It is an established fact that scientists working in narrower disciplines have to depend mostly on specialised journals devoted to their subjects which are ofcourse very few in number, in most cases. Journal of Root Crops occupied the first position with 63 citations, indicating a high self-citation rate. The second position is for the Journal of the American Society for Horticultural Science. It is seen from the ranked list of journals that almost all the journals are related to the field of agriculture in varying degrees and majority of them are available at the library of the Central Tuber Crops Research Institute. (56 out of the first 66 journals)

Use of secondary periodicals

Scientists made use of secondary periodicals to a considerable extent for their information requirements and quoted them as references in the respective papers. This was so, especially in the case of older references, actual copies of which were not available.

Chronology of Journal citations

It was seen that most of the journal citations belonged to the period

1961-70 (27.76%) and a decreasing trend has been noticed in the number of citations from the period 1971 onwards.

Geographical distribution of journal citations

Geographical analysis of the journal citations showed that 28.32% were from USA, 27.46% from India and 20.52% from UK. Thus the literature from these countries accounted for more than 76% of the journal citations. Several other countries engaged in the cultivation and research in the area of tuber crops were also represented.

10. CONCLUSIONS

The analytical study of the papers and the references therein published in the Journal of Root Crops could reveal many findings carrying wide implications to the research policy and the collection development. These findings have thrown much light on the variety of information sources on which the scientists depend on, which in turn help to provide better information services to the scientists concerned and also in improving the existing services. Important conclusions arrived out of the present study are briefly stated below.

As research in tuber crops is mainly centered around the two major crops, viz. Cassava and sweet potato, other crops are not getting due attention and this has resulted in the less number of publications on these crops. Other tuber crops like aroids, yams and minor tubers do deserve due consideration in the context of research and improvement activities.

In addition to the crop production and crop improvement aspects, which constitute the main stay of tuber crops research, it has to diversify more to such areas such as crop utilization and crop biotechnology. More over, extension activities should also be strengthened to popularise these crops.

Journals form the chief source of information to scientists working in the field of tuber crops. This finding supports the acquisition policy which is being followed in the CTCRI Library. The preference given for building an excellent journal collection in the field of agriculture and related aspects is fully justified. This trend has to be continued as this will be highly useful to the scientists as they are found to use more journals than books and other publications.

In the case of collection development, secondary periodicals and other publications like Annual Reports, Symposia//Workshops, Proceedings etc. should be given due representation.

Considering the fact that majority of the contributors belong to CTCRI and that their citation pattern lags behind by about 10 years, there is an urgent need to provide online or CD-ROM search facilities of major data bases in the field. In this context, it would be : ideal to provide CD-ROM search facilities in the CTCRI Library. As many of the databases on agriculture like AGRIS, CAB Abstracts etc. are now available on CD-ROM. The greatest advantage of a CD-ROM database compared to on-line that the more we use, the less will be the cost. Apart from this, to provide a global current information, access to national/international network also need to be explored.

REFERENCES:

1. Advances in Horticulture. Vol. 8. Tuber Crops. New Delhi: Malhotra Publishing House, 1994, 714p.
2. Begum, K.J. and Sami, L.K: Research collaboration in agricultural science. International Library Review 20, 1988.
3. Begum, K.J. and Rajendran, N: Research collaboration in Zoological science. IASLIC Bulletin 35(2), 1980.
4. Britain, J.M: Information needs and application of the results of user studies. In Debons and Cameron, ed. Perspectives in information sciences. 1975.
5. Central Tuber Crops Research Institute, Sreekaryam. Annual report. 1994.
6. Chandrakumarna Nair, M: Information needs and use pattern of agricultural economists. M.L.I.Sc dissertation. Trivandrum University of Kerala, 1985. 180p.
7. Devarajan, G: User's approach to information in libraries. New Delhi: Ess Ess Publications, 1989.
8. Egghe, L: Methodological aspects of bibliometrics. Library Science 143, 1964: pp. 882-884.
9. FAO State of Food and Agriculture 1994. 357p.
10. Garfield, Eugene: Citation indexing-its theory and application in science, technology and humanities. New York: Wiley, 1979 p 274.
11. Garg, K.C. and Dutt, B: Bibliometrics of agricultural research in India. Quarterly Bulletin of the International association of Agricultural Information Specialists. 37(3), 1992: pp. 133-137.
12. Humayoon Kabir,s: Citations in Indian agricultural research-: an analysis. Annals of Library Science and Documentation 37(3), 1990: pp. 102-109.

13. Kay, Daisy E: Root Crops Ed 2. London: TDRI, 1987. p. 380.

14. Line, Maurice B: The "half-life" of periodical literature: apparent and real obsolescence. Journal of Documentation 26(1), 1970: pp. 46-52.

15. Line, Maurice B: Practical interpretation of citations and library use studies. College and Research Libraries. September 1975: pp. 393-396.

16. Maheswarappa, B.S. and Ningoji, M.M: Growth of literature in the field of applied sciences in India (1965-1989). International Information Communication and Education 12(2), 1993: pp. 191-200.

17. Maheswarappa, B.S. and Savadatti, S.G: Authorship pattern and collaborative research in plant breeding (1934-1989). IASLIC Bulletin 35(3), 1990: pp. 119-124.

18. Maheswarappa, B.S. and Prakash, B.P: Literature use pattern by researchers in the field of Botany. Journal of Library and Information 7(1), 1982: pp. 15-32.

19. National Commission On Agriculture Report. Pt. VI. 1976. p. 494.

20. Onuwueme, I.C: The tropical tuber crops. London: Woley, 1978: p. 234.

21. Paisley, W.J: Information needs and uses. Annual Review of Information Science and Technology Vol. 3, 1968: pp. 1-30.

22. Sengupta, I. N: Bibliometrics and identification of core periodicals. Herald of Library Science 29(3-4), 1990: pp. 226-247.

23. Subbaiah, R: Conceptual frame work for an Indian National Agricultural Information System (INAGRIS). University of Kerala. Ph.D. Thesis. 1985.

24. Subramanyam, K. and Stephens, Elsie M: Research collaboration and funding in Biochemistry and Chemical Engineering. International Forum of Information and Documentation 7, 1982: p. 26.

25. Subramanyam, K: Bibliometric studies of research collaboration. Journal of Information Science 6, 1983: pp. 33-38.

26. Swaminathan, M.S: Changing nature of food security challenge: Implications for agricultural research and policy. Washington: CGIAR, 1990. p. 280.

27. Whitney, G: Pattern of authorship in major bibliographic databases, Scientometrics 26(2), 1993: pp. 275-292.

12

SCHOLARLY LITERATURE IN MALAYALAM: A STUDY BASED ON THE BIBLIOMETRIC ANALYSIS OF THE JOURNAL 'VIJNANAKAIRALI'

APARNA P R

0. INTRODUCTION

There cannot be a literature without a language, but there can be language without literature. Literary works in a language play an important role in helping us to learn the history of a language. The relationship between the literary language and the language used in everyday speech is like the relationship between a flower and the plant.

Malayalam is one of the four major Dravidian languages of South India, in terms of number of persons speaking it and its literary importance. The other three languages are Tamil, Kannada and Telugu. Malayalam developed as a language during the 11th and 12th centuries. The Portuguese who came for trade under the leadership of Vasco Da Gama gradually began to take part in politics also. The foreign invasion and the constant state of conflict between the various native kingdoms resulted in restrictions in the social life of people and hampered the growth of literature also. But the political change helped indirectly in the case of prose, as this helped to develop a prose style for the trade language. The political awareness of the people and the atmosphere of agitations prevailing them, to remove the influence of Tamil from the prose and to develop a pure form of Malayalam.

Vijnanakairali is a periodical brought out by the Kerala Bhasha Institute. Kerala Bhasha Institute was established with the aim of publishing University level books in Malayalam in all branches of modern science. It was also the aim of the Institute to make Malayalam the medium of instruction at University level and to develop Malayalam as a strong medium to propagate modern science among Keralites. The Kerala Bhasha Institute was a project under the Fourth Five Year Plan. Sixty percent of the expenditure involved in this project is met by the State Government and forty percent by the Central Government. Though at first, the aim of the Institute was to produce University level text books in Malayalam, later the Institute was determined as an epochal effort to enrich Malayalam language generally and Malayalam scientific literature especially.

Vijnanakairali started publication in June 1969. The aims of Vijnanakairali are the same as the aims of Institute which are as follows:

1. To develop Malayalam as an effective medium for the popagation of modern science and to develop its strength and wealth.
2. To develop the fruitful interaction between the various Indian languages.
3. To help the local language grow as a medium for social and emotional development.

1. SIGNIFICANCE OF THE STUDY

Scholarly periodicals in Malayalam are few and far between. Most of the periodicals in Malayalam are soley entertainment oriented and are of popular nature. A few like Bhashaposhini, Bhasha Sahiti, Sahityalokam etc. are meant for serious reading. But they confine themselves to the field of literature. Vijnanakairali, the official organ of the Kerala Bhasha Institute (State Institute of Languages) is the only periodical giving equal importance to sciences, social sciences and humanities. The main objective of starting this journal was to enrich Malayalam language with scholarly articles in all fields of knowledge. Another objective of the journal was to develop an array of writers who can present complex subjects in lucid Malayalam. Vijnanakairali has completely succeeded in the fulfilment of both these objectives. The earlier issues of Vijnanakairali bear testimony to this fact. They contain authoritative articles written by experts in the relevant fields. Over the years, the journal has become synonymous with scholarly periodical literature in Malayalam. Started in 1969, Vijnanakairali was celebrating its silver jubilee year in 1994. At this juncture it was considered worthwhile to make bibliometric study of the contents of the journal which has paid yeomen service to the development of Malayalam language and literature. Vijnanakairali is committed to the development of the language so that it can be easily used for higher education purposes. Therefore, an analysis of Vijnanakairali would be helpful to the publishers, students and researchers.

2. ANALYSIS

Vijnanakairali is the only prestigious periodical in Malayalam which gives equal importance to science, social science and humanities subjects. It is one of the few periodicals recognised by Government for the publication of papers by research scholars. To that extent, the articles appearing in this journal remain as the yardstick of the quality and quantity of the scholarly literature in Malayalam in different fields of study. For convenience of study and due to time limitations, only the issues for ten years from 1985-1994 were included for the study. Separate entries of articles of these ten years were recorded on 5" x 3" cards and subjected to different analysis. Editorials were excluded from the study.

3. NUMBER OF PAPERS

The number of articles and other items in every issue of Vijnanakairali was analysed. The results are given in Table 1.

Table-1 Numbers of papers covered by Vijnanakairali in each volume

Year	No. of articles	No. of short Communications	Total No. of Contributions
1985 (Vol. 16)	156 (91.76%)	14 (8.23%)	170
1989 (Vol. 17)	108 (98.18%)	2 (1.825)	110
1987 (Vol. 18)	111 (99.10%)	1 (.89%)	112
1988 (Vol. 19)	115 (95.79%)	5 (.42%)	120
1989 (Vol. 20)	124 (99.2%)	1 (.8%)	125
1990 (Vol. 21)	150 (100%)	- -	150
1991 (Vol. 22)	137 (97.85%)	3 (2.14%)	140
1992 (Vol. 23)	152 (99.41%)	1 (.62%)	170
1993 (Vol. 24)	152 (99.34%)	1 (.66%)	153
1994 (Vol. 25)	172 (98.85%)	2 (1.15%)	174
Total	1394	30	1424

The total number of contributions vary from 110 in 1986 to 174 in 1994. There is no consistent growth in the numbers of items covered.

As to the length of the articles, the present editor explained that there is no limitation on the length of articles to be included. If an article is too lengthy, it is published in two or more volumes.

The term 'short communications' is used to designate suippets which are included in each issue. The analysis shows that there is no fixed policy of including such items. They are incorporated subject to the availability of space in each issue.

4. TRANSLATIONS

Vijnanakairali also publishes Malayalam translations of articles published in other languages. The year-wise distribution of translations during the period of study (1985-1994) is shown in the following table.

Table-2 Year-wise distribution of translation

Year	Total No. of articles	No. of trans-lations	Percentage %
1985	170	5	17.24
1986	110	1	3.45
1987	112	4	13.79
1988	120	1	3.45
1989	125	4	13.79
1990	150	-	- -
1991	140	-	- -
1992	170	7	24.13
1993	153	2	6.89
1994	174	5	17.24
Total		29	100

The share of translations is very less.

Language of the translations

An attempt was made to find out the languages from which translations were made and the results are tabulated in Table 3.

Table-3 Language-wise distribution of translation

Languages	No. of translations	Percentage
English	9	31.03
Hindi	7	24.14
Tamil	4	13.79
Kannada	3	10.34
Bengali	1	3.45
Gujarat	1	3.45
Punjabi	1	3.45
Russian	1	3.45
Sanskrit	1	3.45
Urdu	1	3.45
Total	29	100

Analysis shows that English and Hindi are the languages from which most translations were made. While 31.03% of them were made from English, 24.14% were made from Hindi. About 80% of the translations were from four languages - English, Hindi, Tamil and Kannada.

Subject-wise distribution of translations

Vijnanakairali, being a journal which covers almost all subjects, further analysis was carried out to find out the major subject field in which most of the translations were made. Details of the analysis are shown in Table 4.

Table-4 Subject-wise distribution of translations

Subject	No. of translations	Percentage
Literature	15	51.72
Biology	3	10.34
Eduction	3	10.34
Linguistics	3	10.34
Agriculture	1	3.45
Fine Arts	1	3.45
Physics	1	3.45

Politics	1	3.45
Technology	1	3.45
Total	29	100

Out of the 29 articles translated, 15 (51.72%) were in the fields of literature. Only 6 (20.69%) belonged to science and technology.

Pattern of Authorship

It is found that multiple authorship is more common in science subjects than in social sciences and humanities. The entries of articles were examined to find if single authorship was more prevalent. It is seen that single authorship is more prevalent as is seen from Table 5.

Only 15 papers were found to have multiple authorship out of the 1424 papers examined. Out of these 6 papers were published in 1994. In 1993 the corresponding number was 3. This fact is suggestive of the trend towards multiple authorship. Corporate authorship was conspicuous by its absence.

Main contributors in the Journal

An attempt was made to locate the prolific writers in Malayalam in each subject. Authors who had contributed atleast 10 papers in 10 years ie. those who have written an average of one paper in a year were considered. The names of authors, the number of contributions and the subjects in which they have contributed are given in Table 6.

Table-6 Main contributors in the journal

Name	No. of contribution	Subject
T.R. Sreenivas	19	Philosophy, Literature, Science, Biography
V.K. Abdullah	17	Politics, Education, History, Journalism, Communication
Dr. A.C. Fernandes	17	Medicine, Biology
Prof. K.A. Vasukuttan	16	Forestry, Zoology
N.R. Gopinatha Pillai	13	Linguistics
Dr. K.K.N. Kurup	12	History

Table-5

No. of authors	Year 1985		1986		1987		1988		1989		1990		1991		1992		1993		1994	
	No. of papers	%	No. of papers	%	No. of papers	%	No. of papers	%	No. of papers	%	No. of papers	%	No. of papers	%	No. of papers	%	No. of papers	%	No. of papers	%
1	109		110	100	111	120	120	100	123		149	99.33	140	100	169		150	98.03	168	
2 and above	1		-	-	1	-	-		2		1	.66	-	-	1		3	1.96	6	
Total	170		110		112		120		125		150		140		170		153		174	

Dr. R. Mahadevan Nair	12	History, Archeology
Prof. K.N. Ramakrishnan	12	Medicine, Botany
K. Sivasankaran Nair	12	History
Dr. D. Benjamin	11	Linguistics
U. Chandrahasan	11	Physics
Prof. C. Narayanan	11	Polytics, Economics
M.G. Sasibhooshan	11	History, Folklore
Dr. N.M. Sreedharan Nair	11	Physics, Science
S. Vijaykumar	11	Astronomy Linguistics
P.T. Bhaskara Panicar	10	Economics, History, Education, Science
K.M. Govi	10	Library Science
M. Sreenath	10	Linguistics
Total	237	

Sri. T.R. Sreenivas is the most prolific author with 20 contributions in the field of philosophy, science, literature and biography. Sri. V.K. Abdullah and Dr. A.C. Fernandes come next with 17 contributions.

Institutional affiliation of authors

It is a matter of common experience that persons working in academic institutions contribute more than those working in other institutions. In order to find out whether this was true of Malayalam writers, the institutional affiliation of authors was analysed. Almost all the articles contained the address of authors. For convenience, the institutions were broadly divided into colleges, Universities, Kerala Bhasha Institute and other institutions and those without any institutional affiliation. The addresses of 36 authors were not given. The details of the analysis are shown in Table 7.

Table-7 Institution-wise distribution of authors

Type of Institution	No. of authors	Percentage
College	124	19.31
University	202	31.64
Bhasha Institute	13	2.02

Non-academic Institutions	168	26.16
Others (without institutional affiliation)	99	15.42
Addresses not known	36	5.6
Total	642	100

About one-third of the authors belonged to universities. Second come authors working in Non-academic institutions and then come authors working in colleges. Only 13 authors belonging to Bhasha Institute have written in the periodical. This is a bit surprising considering Vijnanakairali is the official organ of the Institute.

Geographical distribution of authors

An attempt was made to study the geographical distribution of the authors. Most of the authors come from Thiruvananthapuram district followed by Kozhikode and Ernakulam districts respectively as can be seen from Table 8.

Table-8 Geographical distribution of authors

District	No. of authors
Thiruvananthapuram	346
Kozhikode	70
Ernakulam	40
Thirssur	36
Kottayam	25
Alappuzha	13
Kannur	9
Kollam	9
Palakkad	8
Kasargode	7
Malappuram	4
Pathanamthitta	4
Wynad	2
Outside Kerala	35
Not known	34
Total	642

More than half of the authors come from Thiruvananthapuram district. Wynad district gets the least representation. There are some authors from outside Kerala as well as from abroad. Their distribution is shown below.

Table 9 Geographical distribution of authors outside Kerala

State	No. of authors
Andhra Pradesh	2
Bihar	1
Goa	1
Gujarat	1
Karnataka	3
Maharashtra	4
New Delhi	5
Pondicherry	1
Tamil Nadu	7
Uttar Pradesh	2
West Bengal	1
Total	28

Most of the authors from outside Kerala come from Tamil Nadu. New Delhi comes next, followed by Maharashtra.

The following is the distribution of authors from outside India.

Table 10 Distribution of authors outside India

Country	No. of authors
U.S.A.	3
Libya	1
Russia	1
Total	5

The contributions of authors from outside India is very less. Out of a total of 642 authors, only 5 are from outside India. Of these, 3 are from USA and one each from Libya and Russia. Only one article is a translation. The rest are in Malayalam.

Sex-wise distribution of authors

An attempt was made to study the proportion of male and female contributors.

Table-11 Sex-wise distribution of authors

Sex	No. of authors
Male	572 (89.09%)
Female	70 (10.9%)
Total	642

Out of a total of 642 authors, only 70 (10.9%) were female, the rest 572 (89.09%) were male. Sri. T.R. Sreenivas is the leading male contributor with 19 contributions and Smt. Aishath Suhra, Smt. K.K. Indira and Dr. B. Prabha are the leading female authors with 4 contributions each.

Subject-wise distribution of papers

Vijnanakairali claims to give equal importance to science, social science and humanities subjects. An attempt was made to find out the number of contributions in each subject. The results are tabulated in Table 12.

Table-12 Subject-wise distribution of papers

Subject	No. of papers	Total
Science and Technology		
Biology	89	
Medicine	75	
Engineering & Technology	58	
Agriculture	41	
Mathematics	37	
Physics	32	
Science	26	
Chemistry	12	
Communication	8	
Geology	5	
		383 (26.85%)

Social Sciences		
History	110	
Economics	63	
Education	58	
Politics	26	
Philosophy	21	
Sociology	20	
Geography	19	
Social Science	7	
Psychology	5	
		329 (23.54%)
Humanities		
Literature	178	
Linguistics	138	
Fine Arts	83	
Biography	71	
Languages	24	
Useful Arts	24	
Arthopology	20	
Library Science	19	
Journalism	6	
Astrology	5	
Religion	5	
		573 (41.25%)
Book Reviews	92	
Short Communications	30	
Reports	10	
Generalia Bibliography	7	
		139
Total		1424

The majority of contributions were in the field of humanities (39.24%) followed by science and technology (26.85%) and social sciences (23.54%). Most of the contributions in Humanities were in the field of literature followed by Linguistics and Fine Arts. In Science and Technology, the majority of contributions were in the field of Biology, followed by Medicine and Technology. History had the first place in social sciences followed by Economics and Education.

5. CITATIONS

A study was also made of the citations given in papers. Some of the references were explanations or clarifications of terms used in the text. Those have not been taken into account while counting the number of citations.

Table-13 Distribution of Citations

Year	Total No. of papers	No. of papers having citation	Percentage	Total No. of citations
1985	170	31	19.02	229
1986	110	15	13.64	190
1987	112	8	7.47	100
1988	120	24	20.16	304
1989	125	20	17.24	250
1990	150	10	6.66	120
1991	140	4	2.87	54
1992	170	11	6.79	119
1993	153	18	11.76	188
1994	174	15	8.82	160
Total	1424	156		1714

The number of papers having citations is very less compared to the total number of papers. The percentage of papers having citations has never been more than 20% of the total number of papers. There is non-consistent growth or regression in the number of papers having citations and the total number of citations.

6. CONCLUSION

Vijnanakairali sets no restriction on the length for the articles to be published it would be helpful to publish more articles in an issue if an acceptable length is prescribed for the contributions. Vijnanakairali needs to give more representation to social subjects like Kerala Politics, Sociology etc. Among the humanities subjects Hinduism, Christianity, Individual Indian languages other than Malayalam and Sanskrit etc. also have not appeared during the last ten years. Vijnanakairali should give more attention to these topics. The percentage of translations published in Vijnanakairali is very less. It would be helpful to the readers if relevant articles in particular fields are included, especially to cover areas in which local talent is not available.

Various research institutions in the state have only very poor representation in the journal. Similarly, very far of the large number of colleges in Kerala are represented. The participation of such authors should be solicited so that the coverage of Vijnanakairali can be enhanced. Similarly the cooperation of more authors from the northern districts of Kerala should be enlisted to increase the popularity of the journal.

Citations were found to be appended to only a few articles. But in Malayalam, some of the articles in Vijnanakairali may be the only source of information on certain topics. It is quite likely that the readers might want to have further reading materials on the topic. Not only such readers, but others would also be benefited if Vijnanakairali insists on citations. This would also add to the authority of the contributions and thereby the prestige of the journal would also increase. An attractive lay out would enhance the marketability of the journal and readability would also increase. It is imperative therefore, that the overall presentation should be made attractive by using modern technology.

REFERENCES:

1. Brooks, BC : Theory of Bradford Laws. Jl. of Doc. 33, 1977.
2. Chummar, T M : 'Bhasha Gadya Sahitya Charithram'. Kottayam, the Author, 1969.
3. George, KM : History of literature through movements. Kottayam, SPCS, 1958.
4. Parameswara, Ulloors : 'Kerala Sahitya Charithram'. Vol.A, Trivandrum, Travancore University, 1964.

5. Rajan Pillai, CV : Malayalam Periodical literature till 1925: A Comprehensive Survey and bibliographical analysis (Ph.D. thesis) University of Kerala, 1986.

6. Sam, N. : 'Malayalathile Sahitya Lekhanangal (Ph.D thesis) University of Kerala, 1982.

13

CITATION CHARACTERISTICS OF INDIAN JOURNAL OF SERICULTURE 1988 - 1992

DR. HUMAYOON KABIR.S &
RAVI B V

0. INTRODUCTION

Silk, the queen of textiles known for its elegance and salubrious nature is being produced from time immemorial. Sericulture is playing a key role in the rural economy of India with its low investment nature and short gestation period and quick and high economic returns in instalments. As a result this agrobased cottage industry has assumed the position of main profession from subsidiary occupation. It provides employment to 6.25 million people (upto 1990) in India. Therefore study and research is of utmost importance in this field. Bibliometric studies reveal the publishing and literature patterns in a field. Plenty of studies have been conducted in a wide variety of subjects. But hardly there are studies on sericulture literature, particularly in the Indian context studies are almost nil. Thus the need of the present study.

1. SCOPE

This study is confined to the citations appended at the end of the articles only. Foot notes and citations of short communications are excluded.

2. OBJECTIVES

The objectives of this study are:

a) To access the number of citations in Indian Journal of Sericulture

b) To identify the different sources of information and their dispersion

c) To identify the subjectwise distribution of citations

d) To identify the languagewise distribution of citations

e) To identify the authorship pattern of the cited articles

f) To find out the proportion of multi-authored vs single authored articles

g) To identify the core and essential journals in the field

h) To identify the Indian journals used by the Indian Sericulture scientists for their research work.

i) To find out whether Indian sericulture scientists use more of foreign or Indian literature

j) To identify the geographical distribution of journals

k) To aid the researchers in the field to be informed about the important periodicals, productive subjects, areas, languages and other trends.

3. LIMITATIONS

The period of coverage is only for five years (ie. 1988-1992). The collection, processing and analysing of the data is carried out manually which may affect the accuracy of the result. Only a single source journal has been analysed for the present study.

4. METHODOLOGY

All the issues of the journal from 1988-1992 were examined. The data relating to all the references appended to the articles during the period have been collected and tabulated. This study has counted five years of data. Information was recorded in slips of 5" x 3" size. After collecting, these were analysed and tabulated for making observations.

5. SOURCE OF DATA

In this study Indian Journal of Sericulture (started in 1962), a publication of the Central Silk Board, Bangalore has been selected as the source journal. This journal is half-yearly and is an important Indian journal in the field of Sericulture

6. ANALYSIS OF DATA

6.1 Citation structure

The collected data in the present study consisted of a total number of 2,254 citations. The total number of articles scanned were 250. The average citation per article was 9.2 tabulated in Table 1.

Table-1 Average Citation Per Article

(a) Article with citations	(b) Article without citations	(c) Total No. of citations	(d) Total No. of citations	(e) Average citation per article (d/a)
245	5	250	2254	9.2

The above table represents a diagrammatic sketch of the total citation structure of the articles of Indian Journal of Sericulture. There were 5 articles without citations. Total citations in all the periodical articles are 2254 and the average citation is 9.2.

6.2 *Formwise distribution*

Table No. 2 gives a clear picture of the different bibliographic forms of literature.

From table 2 it is evident that the major source of information used by the Indian Sericulture Scientists is the journal articles, as it occupies the first position. The total number of citations on journal articles were 1,473 ie. 65.35% of the total citations. The next position was occupied by books with 438 citations forming 19.43%. Books were followed by Institutional publications such as Bulletins and reports with a total of 146 citations (6.46%). Proceedings of Conferences and theses/ Dissertations are having 132 and 43 ie. 5.86% and 1.91% respectively. Others mentioned in the table includes patents, monographs, tables, abstracts, pamphlets, personnel communication etc. which comes to 22 citations ie. 0.98% of the total.

It can be inferred from the study that to have access to maximum volume of information in the field of Sericulture, the journals are the prime channel of information.

Reports and Proceedings are of equal importance because both are having a percentage of 6.45% and 5.86% respectively.

Journal articles together with books cover nearly 85% (exactly 84.78%) of the total citations and therefore these two bibliographic forms are important in the field.

6.3 *Subjectwise Distribution of Citations*

In order to determine the subjects discussed in the citations, the

Table-2 Formwise Distribution of Citations

Sl. No.	Forms of Literature	Rank	Citations 1988	1989	1990	1991	1992	Total	Cumulative	Percentage	Cumulative Percentage
1.	Journal Articles	I	187	352	370	381	263	1473	1473	65.35	65.35
2.	Books	II	75	102	110	69	82	438	1911	19.43	84.78
3.	Bulletins/ Reports	III	30	37	48	14	17	146	2957	6.46	91.25
4.	Proceedings of Symposium Seminar Workshop etc.	IV	18	25	20	27	15	132	2100	5.86	97.11
5.	Thesis/ Dissertations	V	2	13	47	4	4	43	2232	1.91	99.02
6.	Others	VI	4	3	5	8	3	22	2254	0.98	100.00
	Total		316	532	623	423	383	2254		100.00	

subjects were classified according to standard divisions. The citations published in the Indian Journal of Sericulture have been analysed on the basis of their subjects are given in table 3.

Subjectwise distribution of citations as given in table 3 showed that Silkworm physiology occupied the first rank with 649 citations (28.79% of the total). This is followed by Mulberry Propagation, Silkworm Pathology, Silkworm Breeding and Genetics with 231 (10.25%), 205 (9.09%) and 198 (8.78%) citations respectively. The above mentioned five subject divisions altogether occupies more than 70% of the literature. The rest 9 subjects are having citation around 30% only. The miscellaneous subjects includes Zoology, Agricultural Science, Biology, Pure Sciences, Botany, Horticulture etc. which shows the dependence of Sericulture Scientists for information in other subjects also.

On the whole it can be perceived that Indian research in Sericulture is concentrated in Silkworm Physiology, Silkworm Pest Management, Mulberry Propagation and Silkworm Pathology.

6.4 Languagewise Distribution

It is great significance for researcher's to know the language in which their area of specialisation is published. The analysis of the languages used to transmit the subject literature is useful not only to understand the coverage patterns, but also gives an indication of the foreign language problem likely to be faced by the users. For example those who want to learn the subject Ayurveda should be through with Sanskrit language, because lionshare of communications are in Sanskrit. Therefore in this study, in order to analyse the medium of communication of Indian Sericulture literature, the language of the publications were noted and analysed. The results are presented in Table 4.

Table-4 Language-wise Distribution of Citations

Sl. No.	Language	Rank	No. of citations	Cumulative-cotetions	Percentage	Cumulative percentage
1	English	I	1452	1452	64.42	64.42
2	Japanese	II	225	1677	9.98	74.40

3	German	III	202	1879	8.96	83.36
4	Russian	IV	180	2059	7.99	91.35
5	French	V	90	2149	3.99	95.34
6	Italian	VI	63	2212	2.80	98.14
7	Chinese	VII	25	2237	1.11	99.25
8	Czech.	VIII	8	2253	0.35	99.60
9	Korean	X	8	2253	0.35	99.95
10	Polish	IX	1	2254	0.04	99.99
	Total		2254		99.99	

Table No. 4 shows the distribution of citations according to language during the period of study. As the table shows that out of a total 2,254 citations, 1452 of them were in English language forming 64.42% of the total, while all other languages accounted to 35.58%. The second place is occupied by Japanese language with 225 (9.98%) citations. The third rank goes to German with 202 ie. 8.96% of the total. This is followed by Russian, French, Italian, Chinese, Czech, Korean and Polish with 180 (7.99%), 90 (3.99%), 63 (2.80%), 25 (1.11%), 8 (0.35%), 8 (0.35%) and 1 (0.04%) respectively.

Based on the information provided in table 4, it is evident that English language occupies Ist rank with around 65% of the total and is the dominant language. This trend has been observed in other science subjects also. Authors from non-english speaking countries and particularly from the developing countries are writing in English in order to have the opportunity of much greater readership for their literature. Therefore they either publish their literature from publications of English speaking countries or publish in English language from their own countries. The existence of publishing in English language from their own countries is more now a days. So from this study it is found that English is the dominant language, not only in natural sciences but also in an applied science area like Sericulture. Hence English language document on sericulture would be very helpful for researcher's in the subject area and should be strengthened in the documentation centres.

6.5 Authorship Pattern

The characteristics of any subject literature include not only the

Table-3 Subjectwise Breakup of Citations

Sl. No.	Subject	Rank	Citations 1988	1989	1990	1991	1992	Total	Cumu-lative	Percentage	Cumu-lative Percentage
1	Silkworm Physiology	I	103	239	84	197	26	649	28.79	28.79	
2	Silkworm Pest Management	II	17	38	154	54	35	298	947	13.22	42.01
3	Silkworm Mulberry Propagation	III	13	26	100	18	74	231	1178	10.25	52.26
4	Silkworm Pathology	IV	21	41	56	32	55	205	1383	9.09	61.35
5	Silkworm Breeding and Genetics	V	20	36	67	41	34	198	1581	8.78	70.13
6	Mulberry Breeding and Genetics	VI	54	23	34	13	31	155	1736	6.88	77.01

Sl. No.	Subject	Rank	Citations 1988	1989	1990	1991	1992	Total	Cumu-lative	Percent age	Cumu-lative Percentage
7	Ericulture	VII	40	37	47	4	24	152	1989	6.74	83.71
8	Mulberry Pathology	VIII	16	14	40	10	43	123	2011	5.46	89.21
9	Pre & Post Silk Technology	IX	16	-	16	28	17	77	2988	3.42	92.63
10	Tasa Culture	X	9	18	-	29	8	62	2150	2.75	95.38
11	Mull berry Physiology	XI	-	-	7	20	19	46	2196	2.04	97.42
12	Muga Culture	XII	-	14	-	-	17	31	2227	1.38	98.80
13	Mul berry Pest Management	XIII	-	-	4	-	10	14	2241	0.62	99.42
14	Miscellaneous	XIV	-	11	-	2	-	13	2254	2.59	100.00
	Total		309	497	509	449	391	2254		100.00	

basic publishing patterns but that of the authors themselves. So the authors were analysed to determine the percentage of single, double, three and more than three authors. It is to be noted here that the authorship pattern has been studied by analysing the journal article citations only. In order to have a clear picture the results of the analysis of authors are presented in Table 5.

It is identified from table 5 that Double Author in Indian Sericulture is in vogue with 469 forming 31.84% of the total number of citations in the field. Three author contributions rank next highest with 431 forming 29.26% of the total. This is followed by single author which is 419 in number with 28.44% of the total and the last rank was occupied by more than three author combination pattern which is 154 in number and constitutes 10.45% of the total. Thus in the field of sericulture collaborative research or team research is playing important role in communication of literature.

6.6 *Proportion of Multi-authored Vs Single-authored Articles*

Analysis of data in table 6 shows that multiple author is predominant. Out of the total 1473 articles 1054 articles (ie. 71.55%) are multiple authored. The ratio between Multi-authored paper Vs Single-authored is 2.51:1.

Table-6 Multiple Author vs Single Authored Articles

Single Author		Multiple Author		
No. of Articles	Percentage	No. of Articles	Percentage	Total No. of Articles
419	28.45	1054	71.55	1473

The low incidence of single authorship and high level of multiple authorship is characteristic of the science subjects. Clarke (30) reported in 1964 that in physical sciences the percentage of multiauthored papers is reported to be in the range of 167 to 83.

6.7 *Journalwise Distribution (Both from India and Abroad)*

Since the journal articles formed the major sources of information, for the Indian Sericulture Scientists, these were further analysed to find out the core and essential journals in the discipline. A rank list was prepared and is given in Table 7.

Table-5 Pattern of Authorship

Sl. No.	Authorship	Rank No.	Year 1988	1989	1990	1991	1992	Total articles	Cumu lative articles	Percent age	Cumu-lative Percentage
1	Double Author	1	104	92	114	103	56	469	469	31.84	31.84
2	Three Author	2	61	103	127	88	52	431	900	29.26	61.10
3	Single Author	3	70	76	98	112	63	419	1319	28.44	89.54
4	More than Three Author	4	28	30	31	39	26	154	1473	10.45	99.99
	Total		263	301	370	342	197	1473		99.99	

Table-7 Rank list of Journals (Both India and Abroad)

Sl. No.	Name of the Journal	Country	Rank No.	1988	1989	1990	1991	1992	Total	Cumuative No. of articles	Percentage	Cumulative percentage
1	2	3	4	5	6	7	8	9	10	11	12	13
1.	Indian Journal of Sericulture	India	I	30	40	79	58	25	232	232	15.75	15.75
2.	Journal of Sericulture Science	Japan	II	12	31	18	20	34	115	347	7.81	23.56
3.	Journal of Insect Physiology	UK	III	16	13	8	21	18	76	423	5.16	28.72
4.	Bulletin of Sericulture Experimental Station	Japan	IV	2	21	10	16	17	66	489	4.48	33.20
5.	Journal Current Science	US	V	3	7	26	10	8	54	543	3.67	36.87
6.	Sericoiogia	France	VI	9	7	16	15	5	52	595	3.53	40.40
7.	Indian Silk	India	VII	5	14	16	3	6	44	639	2.99	43.39
8.	Journal of Biochemistry	Japan	VIII	8	15	2	5	4	34	673	2.30	45.69
9.	Journal of Biological Chemistry	US	VIII	5	3	7	14	5	34	707	2.30	47.99

Sl. No.	Name of the Journal	Country	Rank No.	1988	1989	1990	1991	1992	Total	Cumu-ative No. of arti-cles	Percent-age	Cumula-tive percent-age
10.	Annais, Entomological Society of America	US	IX	3	7	16	4	-	30	737	2.04	50.03
11.	Annais Review of Entomology	US	X	5	4	4	4	3	20	757	1.36	51.39
12.	Indian Journal of Agricultural Science	India	X	4	2	7	3	4	20	777	1.36	52.75
13.	Journal of Invertebrate Pathology	US	XI	1	8	3	1	4	17	794	1.15	53.9
14.	Journal of Plant Physiology	GW	XIII	-	1	6	2	7	16	810	1.09	54.99
15.	Comparative Bio-Chemistry and Physiology	SE	XIII	3	2	1	2	7	15	825	1.02	56.01
16.	Journal of Economic Entomology	US	XIV	2	3	8	1	-	14	839	0.95	56.96
17.	Entomon	India	XIV	1	-	8	3	2	14	853	0.95	57.91
18.	Indian Journal of Entomology	India	XV	1	1	4	2	4	12	877	0.81	58.72
19.	Journal of Experimental Biology	UK	XV	2	2	1	6	1	12	877	0.81	59.53

Sl. No.	Name of the Journal	Country	Rank No.	1988	1989	1990	1991	1992	Total	Cumulative No. of articles	Percentage	Cumulative percentage
20.	Applied Entomology Zoology	Japan	XV	-	3	1	4	3	12	889	0.81	60.34
21.	Advances in Protein Chemistry	US	XVI	-	11	-	-	-	11	900	0.7561.09	
22.	Phytopathology	GW	XVII	-	-	4	-	6	10	910	0.68	61.77
23.	Canadian Entomology	Canada	XVII	-	1	7	1	1	10	920	0.68	62.45
24.	Z Angew Entomol	GW	XVII	-	6	-	1	3	10	930	0.68	63.13
25.	Z Vergl Physiology	UR	XVIII	-	-	4	4	-	8	938	0.54	63.67
26.	Journal of Silkworm	JA	XVIII	-	5	-	2	1	8	946	0.54	64.21
27.	Science	US	XVIII	1	-	2	4	1	8	954	0.54	64.75
28.	Biol Nauk	UR	XVIII	-	6	2	-	-	8	962	0.54	65.29
29.	Indian Foester	India	XIX	3	1	-	1	2	7	969	0.47	65.76
30.	Clinical Chimica Acts	NE	XIX	5	-	2	-	-	7	976	0.47	66.23
31.	General Comparative Endocrinology	UK	XIX	2	-	-	5	-	7	983	0.47	66.7
32.	Nature	UK	XIX	1	4	-	1	1	7	990	0.47	67.17
33.	Agriculture Biochemistry	UK	XIX	-	4	1	-	2	7	997	0.47	67.64
34.	Indian J. Experimental Biology	India	XIX	1	-	2	1	3	7	1004	0.47	68.11

Sl. No.	Name of the Journal	Country	Rank No.	1988	1989	1990	1991	1992	Total	Cumuative No. of articles	Percentage	Cumulative percentage
35.	J. Morphology	US	XIX	5	-	-	1	1	7	1011	0.47	68.58
36.	Insect Biochemistry	UK	XX	-	3	1	1	1	6	1017	0.41	68.99
37.	Naturwissenscnaften	GW	XX	-	-	-	5	1	6	1023	0.41	69.4
38.	Indian Phytopathology	India	XXI	-	1	2	1	1	5	1028	0.34	69.74
39.	J. Scientific and Industrial Res.	India	XXI	-	2	2	1	-	5	1033	0.34	70.08
40.	Oikos	DENMARK	XXI	-	1	-	4	-	5	1043	0.34	70.42
41.	Sericology	INDIA	XXI	1	-	2	1	1	5	1043	0.34	70.76
42.	Biology Zh. Armenia	UR	XXI	-	4	1	-	-	5	1048	0.34	71.1
43.	Japan J. Applied Entomology Zoology	JAPAN	XXI	-	-	-	1	4	5	1053	0.34	71.44
44.	Indian J. Genetics	INDIA	XXI	1	-	1	1	2	5	1058	0.34	71.78
45.	Indian Farming	INDIA	XXI	-	3	1	1	-	5	1063	0.34	72.12
46.	Biological Bulletin	CHINA	XXI	1	1	2	1	-	5	1068	0.34	72.46
47.	Biokhimiva	RUSSIA	XXI	-	4	1	-	-	5	1073	0.34	72.8
48.	Geobios	INDIA	XXI	-	-	3	-	2	5	1078	0.34	73.14

Sl. No.	Name of the Journal	Country	Rank No.	1988	1989	1990	1991	1992	Total	Cumu-ative No. of arti-cles	Percent-age	Cumula-tive percent-age
49.	Entomology Experimental Application	NE	XXI	-	-	3	1	1	5	1083	0.34	73.48
50.	Heridity	UK	XXI	-	-	5	-	-	5	1098	0.34	73.82
51.	Zpjlinzen Physico	SZ	XXI	-	-	3	-	2	5	1093	0.34	74.16
52.	Naturforsch	GERMANY	XXI	-	-	1	4	-	5	1090	0.34	74.5
53.	J. Vertebrate Zoology	RUSSIA	XXI	1	1	1	2	-	5	1103	0.34	74.84
54.	Advances in Insect Physiology	US	XXI	1	1	1	-	2	5	1100	0.34	75.10
55.	Insect Science & Its Applications	US	XXII	-	-	1	1	2	4	1112	0.27	75.45
56.	Indian J. Biochemistry	INDIA	XXII	-	1	-	1	2	4	1116	0.27	75.72
57.	Indian Agriculture	INDIA	XXII	-	-	1	1	2	4	1120	0.27	75.99
58.	J of Cell Biology	US	XXII	-	3	1	-	-	4	1124	0.27	76.26
59.	Indian J. Plant Physiology	INDIA	XXII	-	-	2	-	2	4	1128	0.27	76.53
60.	J of Clinical Pathology	UK	XXII	-	4	-	-	-	4	1132	0.27	76.8
61.	J of Agriculture Science	UK XXII	-	-	2	-	2	4	1136	0.27	77.07	
62.	Journal of Mysore University	INDIA	XXII	2	-	-	-	2	4	1140	0.27	77.34

Sl. No.	Name of the Journal	Country	Rank No.	1988	1989	1990	1991	1992	Total	Cumu-ative No. of arti-cles	Percent-age	Cumula-tive percent-age
63.	Canadian Journal of Botany	CANADA	XXII	-	-	2	-	2	4	1144	0.27	77.61
64.	Comptes, Renous Academic Science	FRANCE	XXII	-	-	1	3	-	4	1148	0.27	77.88
65.	Cytologia	JAPAN	XXII	-	2	-	-	2	4	1152	0.27	78.15
66.	Caryologia	ITALY	XXII	3	1	-	-	-	4	1156	0.27	78.42
67.	Science & Culture	UK	XXII	1	1	1	1	-	4	1160	0.27	78.69
68.	Italian Agriculture	ITALY	XXII	-	-	1	3	-	4	1164	0.27	78.96
69.	Biochimica-et-Biophysica Acts	NE	XXII	-	3	1	-	-	4	1168	0.27	79.23
70.	Journal of Insect Pathology	US	XXIII	-	1	1	-	1	3	1171	0.20	79.43
71.	Plant & Cell Physiology	JAPAN	XXIII	-	2	-	1	-	3	1174	0.20	79.63
72.	Pesticides	INDIA	XXIII	-	2	-	-	1	3	1177	0.20	79.83
73.	Experentia	SE	XXIII	-	-	-	2	1	3	1180	0.20	80.03
74.	The Philippine Agriculturist	PHILLI-PPINES	XXIII	-	2	-	1	-	3	1183	0.20	80.23

Sl. No.	Name of the Journal	Country	Rank No.	1988	1989	1990	1991	1992	Total	Cumuative No. of articles	Percentage	Cumulative percentage
75.	Agricultural Wastes	UK	XXIV	-	-	-	-	2	2	1185	0.13	80.36
76.	Genen Phaenen	GW	XXIV	-	-	2	-	-	2	1187	0.13	80.49
77.	Monatsh Chem	GW	XXIV	-	-	-	2	-	2	1189	0,13	80.62
78.	Mysore J of Agricultural Science	INDIA	XXIV	-	-	-	1	1	2	1191	0.13	80.75
79.	Mitt Medder Fak Kais, University	JAPAN	XXIV	-	-	-	2	-	2	1193	0.13	80.88
80.	Uosa	GERMANY	XXIV	-	-	-	2	-	2	1195	0.13	81.01
81.	Tea	KENYA	XXIV	-	-	2	-	-	2	1197	0.13	81.14
82.	Zoologiscne, Janrbliecner, Abteilung	GW	XXIV	-	-	-	2	-	2	1199	0.13	81.27
83.	Olcologia	SE	XXIV	1	-	-	1	-	2	1201	0.13	81.4
84.	Oriental Insects	US	XXIV	-	1	1	-	-	2	1203	0.13	81.53
85.	Hilgardia	US	XXIV	-	-	2	-	-	2	1205	0.13	81.66
86.	Hoppe-Seyiers Z Physiology Chemistry	RUSSIA	XXIV	-	-	-	2	-	2	1207	0.13	81.79

Sl. No.	Name of the Journal	Country	Rank No.	1988	1989	1990	1991	1992	Total	Cumu-ative No. of arti-cles	Percent-age	Cumula-tive percent-age
87.	Embryologia	JAPAN	XXIV	-	1	-	1	-	2	1209	0.13	81.92
88.	Experimental Agriculture	UK	XXIV	-	-	-	1	1	2	1211	0.13	82.05
89.	Experimental Agriculture	UK	XXIV	-	-	1	-	1	2	2213	0.13	82.31
90.	Prioroda	RUSSIA	XXIV	-	2	-	-	-	2	1216	0.13	82.31
91.	Pesticides, Biochemistry, Physiology	US	XXIV	2	-	-	-	-	2	1216	0.13	82.44
92.	Botanical Magazine-Tokyo	JAPAN	XXIV	-	1	-	1	-	2	1210	0.13	82.57
93.	Bangladesh J. Zoology	BANGLA-DESH	XXIV	-	-	1	1	-	2	1220	0.13	82.7
94.	Biological Review	CZECHO-SLOVAKIA	XXIV	-	-	1	-	1	2	1222	0.13	82.83
95.	Bulletin, Imperial Kyoto, Seri College	JAPAN	XXIV	-	2	-	-	-	2	1224	0.13	82.96
96.	Botanical Review	US	XXIV	1	-	-	-	1	2	1226	0.13	83:22
97.	Bulletin Entonology Research	UK	XXIV	-	1	-	1	-	2	1228	0.13	83.22
98.	Bulletin, Imperial Sericulture Experiment Station	JAPAN	XXIV	-	-	-	2	-	2	1230	0.13	83.35

Sl. No.	Name of the Journal	Country	Rank No.	1988	1989	1990	1991	1992	Total	Cumulative No. of articles	Percentage	Cumulative percentage
99.	Biochemical Genetics	US	XXIV	-	1	-	-	1	2	1232	0.13	83.48
100.	Journal of Agricultural Research	PAKISTAN	XXIV	-	1	1	-	-	2	1234	0.13	83.61
101.	Kanyo Gijutsu	JAPAN	XXIV	-	2	-	-	-	2	1236	0.13	83.74
102.	Canadian Journal of Zoology	CANADA	XXIV	-	-	2	-	-	2	1238	0.13	83.87
103.	Colemania	FRANCE	XXIV	-	-	2	-	-	2	1240	0.13	84
104.	Crop Science	US	XXIV	-	-	1	1	-	2	1242	0.13	84.13
105.	Chromosoma	GERMANY	XXIV	-	-	1	1	-	2	1244	0.13	84.26
106.	Sobsch Akademy Nauk Gruz USR	RUSSIA	XXIV	1	-	-	-	1	2	1246	0.13	84.39
107.	Archives, Biochemistry and Bio-Physics	FRANCE	XXIV	1	1	-	-	-	2	1248	0.13	84.52
108.	Annal Chemica	FRANCE	XXIV	-	-	1	1	-	2	1250	0.13	84.65
109.	Annal of Botanic Pennici	FINLAND	XXIV	-	-	1	1	-	2	1252	0.13	84.78
110.	Annal Review of Plant Physiology	US	XXIV	-	-	2	-	-	2	1254	0.13	84.91

Sl. No.	Name of the Journal	Country	Rank No.	1988	1989	1990	1991	1992	Total	Cumuative No. of articles	Percentage	Cumulative percentage
111.	Australian Bulletin of Entomology	AUSTRALIA	XXIV	-	-	2	-	-	2	1256	0.13	85.04
112.	Annual review of Phytopathology	US	XXIV	-	-	2	-	-	2	1258	0.13	85.17
113.	Analytical Bio-chemistry	US	XXIV	-	2	-	-	-	2	1260	0.13	85.3
114.	Archives Zoology	ITALY	XXIV	1	1	-	-	-	2	1262	0.13	85.56
115.	Agricultural Research Journal	INDIA	XXIV	-	1	-	-	1	2	1264	0.13	85.56
116.	American Journal of Botany	US	XXIV	-	-	2	-	-	2	1266	0.13	85.69
117.	Acta Entomologica Bohemo slovaca	CHINA	XXIV	-	-	-	2	-	2	1268	0.13	85.82
118.	Acta Sericoigica	JAPAN	XXIV	-	-	-	1	1	2	1270	0.13	85.95
119.	Archives Internationales de physioiogie-et-de-Bio-chemie	BELGIUM	XXIV	-	-	-	1	1	2	1272	0.13	86.08
120.	Analytical Chemistry	US	XXIV	-	-	-	1	1	2	1274	0.13	86.21
121.	Journal of Applied Bacteriology	UK	XXIV	-	2	-	-	-	2	1276	0.13	86.34
122.	Journal of Agricultural Research Quarterly	CHINA	XXIV	-	2	-	-	-	2	1278	0.13	86.47
123.	Journal of Entomological Research	INDIA	XXIV	-	1	-	1	-	2	1280	0.13	86.6

Sl. No.	Name of the Journal	Country	Rank No.	1988	1989	1990	1991	1992	Total	Cumulative No. of articles	Percentage	Cumulative percentage
124.	Journal of Chromatographic Science	NE	XXIV	-	-	-	2	-	2	1282	0.13	86.73
125.	Justus - Lieoigs Annal Chemie	GERMANY	XXIV	-	-	-	2	-	2	1284	0.13	86.86
126.	Journal of Embryology Experiment Development	UK	XXIV	1	1	-	-	-	2	1286	0.13	86.99
127.	Journal of Department Agriculture Kyushu University	JAPAN	XXIV	-	-	1	1	-	2	1288	0.13	87.12
128.	J Faculty Textile Science Technology	NE	XXIV	-	-	2	-	-	2	1290	0.13	87.25
129.	Japan J. Medical Science Biology	JAPAN	XXIV	2	-	-	-	2	1292	0.13	87.38	
130.	Journal of Plant Science Research	IRELAND	XXIV	-	-	2	-	-	2	1294	0.13	87.51
131.	Journal of Experimental Research	UK	XXIV	1	1	-	-	-	2	1296	0.13	87.64
132.	Journal of Zoology	UK	XXIV	1	1	-	-	-	2	1298	0.13	97.77

Sl. No.	Name of the Journal	Country	Rank No.	1988	1989	1990	1991	1992	Total	Cumu-ative No. of arti-cles	Percent-age	Cumula-tive percent-age
133.	J. Agricultural Chemistry & Biological Society	JAPAN	XXIV	-	1	-	1	-	2	1300	0.13	97.9
134.	Japanese Journal of Genetics	JAPAN	XXIV	-	2	-	-	-	2	1302	0.13	88.03
135.	J. of Peotide Protein Research	US	XXIV	-	2	-	-	-	2	1304	0.13	88.16
136.	Journal of Agronomy	US	XXIV	-	2	-	-	-	2	1306	0.13	88.29
137.	Indian Textile Journal	INDIA	XXIV	-	1	-	-	1	2	1308	0.13	88.42
138.	Indian Journal of Agronomy	INDIA	XXIV	-	-	1	-	1	2	1310	0.13	88.55
139.	Scientia Morticulura	CHINA	XXIV	-	-	-	-	2	2	1312	0.13	88.68
140.	Journal of General Virology	UK	XXIV	-	-	-	-	2	2	1314	0.13	88.81
141.	Physical Plantum	DENMARK	XXIV	-	-	-	-	2	2	1316	0.13	88.94

From the table 7. it is clearly evident that Indian Journal of Sericulture (published by Central Silk Board, Bangalore) ranks first with the highest concentration of citations on sericulture. It is having 232 citations forming 15.75% of the total. Indian Journal of Sericulture is followed by Journal of Sericulture Science of Japan with 115 citations and forms 7.81% of the total contributions. It is followed by Journal of Insect Physiology (of U.K.) having 76 citations (5.16%) and occupying the third rank in the list. The fourth journal in the list is Bulletin of Sericulture Experimental Station, Japan. This journal received 66 citations (4.48%).

The above mentioned four journals in the ranklist is having about 1/3rd of the citations ie. these carries approximately 33% (489 citations out of 1473) of citations. Out of 297 journals in the list, the rest 293 (297-4 = 293) journals are having about 67% of citations. This shows the utmost importance of the first four journals and are said to be the core journals in the discipline.

Next in order in the list are Journal of Current Science (USA), Sericologia (France), Indian Silk, Journal of Biochemistry (Japan), Journal of Biological Chemistry (USA) and Annals of Entomological Society of America (USA) occupying 5th, 6th, 7th, 8th and 9th ranks respectively. Now the first 10 journals in the ranklist, sharing 9 ranks, are having about 50% of citations. These peripheral journals eventhough outside the core are also important.

Altogether there are 297 journals identified through this study which contains citations on sericulture. In total these 297 journals share 25 ranks. The last 156 journals in the rank list, occupying rank number 25, each contributing only one article are the least important journals.

The above used rank may be considered as a practical tool to select journals of maximum utility for the Indian Sericulture Scientists. By doing this the librarian can tide over the financial crisis due to the price like of journals and limitations of resources. The ranking of journal provides an approach for inclusion or exclusion of journals to the extent the budget permits. However, the relative importance of a specific journal may change due to shift in R & D work of the institution or emergence of more appropriate journal covering sub-disciplines representing the scientists work. So periodic review of ranking of journals every decade should be beneficial.

6.8 Identification of Indian Journals

Journal list is given in table 7 was further analysed to find out the Indian journals cited by the Indian Sericulture Scientists. A ranklist of Indian journals was prepared and is given in table 8.

Table-8 Rank list of Indian Journals

Sl. No.	Title of the Journal	Rank	No. of citations	% age out of 1473
1	Indian Journal of Sericulture	I	232	15.75
2	Indian Silk	II	44	2.99
3	Indian Journal of Agricultural Science	III	20	1.36
4	Entomon	IV	14	0.95
5	Indian Journal of Entomology	V	12	0.81
6	Indian Forester	VI	07	0.47
7	Indian Journal of Experimental Biology	VI	07	0.47
8	Indian Phytopathology	VII	05	0.34
9	Journal of Scientific & Industrial Research	VII	05	0.34
10	Sericology	VII	05	0.34
11	Indian Journal of Genetics	VII.	05	0.34
12	Indian farming	VII	05	0.34
13	Geobios	VII	05	0.34
14	Indian Journal of Bio-Chemistry	VIII	04	0.27
15	Indian Agriculturist	VIII	04	0.27
16	Indian Journal of Plant Physiology	VIII	04	0.27
17	Journal of Mysore University	VIII	04	0.27
18	Pesticides	IX	03	0.20

19	Mysore Journal of Agricultural Science	X	02	0.13
20	Journal of Entomological Research	X	02	0.13
21	Indian Textile Journal	X	02	0.13
22	Indian Journal of Agronomy	X	02	0.13
23	Memoirs-Department of Agriculture	XI	01	0.07
24	The Forester	XI	01	0.07
25	Krishivarthe	XI	01	0.07
26	Current Research UAS	XI	01	0.07
27	Journal of Indian Institute of Science	XI	01	0.07
28	Journal of Indian Botany	XI	01	0.07
29	Journal of Maharashtra Agricultural Universities	XI	01	0.07
30	Indian Journal of Physiology & Applied Ecology	XI	01	0.07
31	Insect Pests	XI	01	0.07
32	Indian Journal of Genetics & Plant Breeding	XI	01	0.07
33	Journal of Agricultural Research-Punjab Agricultural] University	XI	01	0.07
34	Bio-Research	XI	01	0.07
35	Plant Biochemical Journal	XI	01	0.07
36	Khadi Grama Udyog	XI	01	0.07
37	ICAR Technical Bulletin	XI	01	0.07
	Total		406	27.69

It is found from the study that there are 37 Indian journals in the list. The journals are arranged according to the citations they received. Indian Journal of Sericulture with 232 citations ranks the top. The same journal is also coming to the top position in the common ranklist of Indian & foreign journals also. Out of 1473 citations, the citations of Indian Journals are 406 (27.69%). Out of a total of 297 journals (both Indian and Foreign), only 37 journals are from India, which comes to 12.46%. This clearly indicates that Indian researchers in Sericulture use more foreign literature for their research work than Indian ones.

6.9 *Geographicalwise distribution of journals*

The data regarding periodicals were analysed according to their country of origin and is given in table no. 9.

Table No. 9 gives the geographical distribution of journals in sericulture. This table contains a list of 28 countries which shares 14 ranks, wherein the number of periodicals published from these 28 countries is 297 journals of the total articles published is 1473.

USA ranks first with 62 periodicals forming 20.80% of the 297 journals. These 62 periodicals are having a total contribution of 274 articles forming 18.6% of the total 1473. It is followed by Japan contributing 39 periodicals (13.13%) India ranks next to Japan with 37 periodicals 12.46%). This is followed by UK, Germany and Russia sharing IV, V and VI ranks respectively with their periodical contributions 35 (11.78%), 27 (9.09%) and 20 (6.73%) respectively.

From the above table we can infer that six countries (USA, Japan, India, UK, Germany and Russia) which shares the first six ranks contribute approximately 3/4th of the periodicals (74.06%) published during the stipulated period. Rest 1/4th of the periodicals are from 22 countries (28 - 6 = 22). The last 9 countries in the list contributes only one periodical each. Which is of least importance.

From the above analysis we can conclude that research in sericulture is quite good in USA, because the journal output is much high in this country. It is further inferred that research on sericulture is well organised in countries like USA, Japan, India, UK, Germany, Russia etc. Hence it is clear from this study that USA publishes maximum number of journals in sericulture area.

Table-9 Countrywise Distribution of Periodicals and Articles

Sl.	Country Journals	Rank of Journals	No. of Journals	Cumulative	PercentageNo. Percentage	Cumulative Articles	No. of Articles	Cumulative	Percentage Percentage	Cumulative
1.	United States	I	62	62	20.87	20.87	274	274	18.60	18.60
2.	Japan	II	39	101	13.13	34.00	289	563	19.62	30.22
3.	India	III	37	138	12.46	46.46	409	972	28.80	66.02
4.	United Kingdom	IV	35	173	11.78	58.24	164	1136	11.13	77.15
5.	Germany	V	27	200	9.09	67.33	75	1211	5.09	82.24
6.	Russia	VI	20	220	6.73	74.06	53	1264	3.59	85.83
7.	France	VII	13	233	4.38	78.44	69	1333	4.68	90.51
8.	Italy	VIII	10	243	3.37	81.81	17	1350	1.15	91.66
9.	Netherlands	IX	9	252	3.03	84.84	22	1372	1.49	93.51
10.	Canada	X	5	257	1.68	86.52	18	1390	1.22	94.37
11.	China	X	5	262	1.68	88.82	12	1402	0.81	95.18
12.	Australia	X	5	267	1.68	86.88	6	1408	0.40	95.58
13.	Switzerland	XI	4	271	1.35	91.23	25	1433	1.70	97.28
14.	Denmark	XI	4	275	1.35	92.59	9	1442	0.61	97.89

Sl.	Country Journals	Rank of Journals	No. of Journals	Cumulative	PercentageNo. Percentage	Cumulative Articles	No. of Articles	Cumulative	Percentage Percentage	Cumulative
15.	Pakistan	XII	3	278	1.01	93.59	4	1446	0.27	98.18
16.	Belgium	XII	3	281	1.01	94.80	4	1450	0.27	98.43
17.	South Korea	XII	3	284	1.01	95.81	3	1453	0.20	98.63
18.	Ireland	XIII	2	286	0.67	96.28	3	1456	0.20	97.83
19.	Finland	XIII	2	288	0.67	96.95	3	1459	0.20	99.03
20.	Philippines	XIV	1	289	0.34	97.29	3	1462	0.20	99.23
21.	Kenya	XIV	1	290	0.34	97.63	2	1464	0.14	
22.	Bangladesh	XIV	1	291	0.34	97.97	2	1466	0.14	
23.	Czechoslovakia	XIV	1	292	0.34	98.31	2	1468	0.14	
24.	Argentina	XIV	1	293	0.34	98.65	1	1469	0.07	
25.	South Africa	XIV	1	294	0.34	98.99	1	1470	0.07	
26.	Poland	XIV	1	295	0.34	99.33	1	1471	0.07	
27.	Srilanka	XIV	1	296	0.34	99.67	1	1472	0.07	
28.	Turkey	XIV	1	297	0.34	100.00	1	1473	0.07	
	Total			297	100.00		1473	1473	100.00	

7. CONCLUSION

Average number of citations in Indian Journal of Sericulture is 9.2. The major source of information used by the sericulture scientists are the journal articles. Thus to have maximum volume of information in the field, the journals are the prime channel of information. Among the subjects Silkworm Physiology ranks the top, which shows the thrust given by the researchers in this subject division.

English language was the predominant language. Thus it is the most profusely used and most important language for communication. Double author contributions are maximum with about 1/3rd of the total. Multiple author contributions predominates which shows collaborative research is popular in the field. It is also found that increase in collaboration of authors decreases the productivity. Ratio between multiauthored and single authored articles are 2.51:1. Among the journal citations Indian Journal of Sericulture ranks the top. Further it is found that Indian researchers use more foreign journals for their research work that of Indian journals. The journals they use are from 28 different countries, of which journals from USA ranks the highest.

As mentioned earlier the study of single journal is of limited usefulness. Moreover the data is available only for a period of five years. By increasing the period of study and inclusion of more number of source journals, one may get a fairly comprehensive trend of the citation characteristics of Indian Sericulture literature as a whole.

REFERENCES:

1. Adeniran O R: Bibliometrics of Computer Science literature in Nigeria. International Library Review, 20, 1988, p. 347-359.
2. Aiyepeku W O: The Bradford distribution theory: The compounding of Bradford periodical literature in geography. J. of Documentation. 33, 3, 1977, pp. 210-219.
3. Alabi G A: The citation pattern of Nigerian Scientists. Intl. Library Review, 21, 1, 1989, pp. 129-137.
4. Alafiatayo, Benjamin O: The research literature of agricultural economics. International Library Review. 21, 1989, p. 463-479.
5. Arunachalam, S and Hirannaiah S: Journal preference of Indian researchers in physical and chemical sciences. Proceedings of the Seminar on primary communication in science and technology in India. 4-8 Dec. 1978, pp. 141-148.

6. Asundi, A Y and Humayoon Kabir S. Evolving criteria for identification and selection of core periodicals in a subject : A case study on Indian Horticulture. Library Science with a Slant to Documentation and Information Studies (Accepted).

7. Baker, D: Recent trends in the growth of chemical literature. Chemical Engineering News, 54, 1976, pp. 23-27.

8. Balasubramanian S and Bhaskar P A: Citation analysis in neurology from the point of view of Indian Scientists. Iaslic Bulletin. 29, 1984, pp. 63-67.

9. Begum K J and Rajendra N: Research collaboration in zoological sciences. Iaslic Bulletin. 35, 2, 1990 pp. 79-82.

10. Broadus, R N: Definition of bibliometrics. Scientometrics. 12, 1987, pp. 373-379.

11. Chakraborty R: Citation analysis of petrological literature. Lucknow Librarian, 15, 1983, pp. 99-104.

12. Cline Gloria Stark. A bibliometric study of two selected journals in Library Science, 1940-1974. Ph.D. Dissertation, University of Southern California.

13. Craig G R: Communication in geology. Scottish Journal of Geology. 5, 1969, pp. 305-321.

14. Donohue Joseph C. Understanding scientific literature : A bibliometric approach. 1973. Cambridge, MIT Press.

15. Fenner P and Lloyd J. Serials important to geologists. Geotimes. 15, 1970. pp. 14-19.

16. Gupta B M. Citation analysis and its applications : A review. Herald of Library Science. 18, 1-2, 1979, p. 88.

17. Gupta D K and Gupta S. Periodical literature of petroleum geology : A citation analysis study to determine obsolescence factors and patterns. Iaslic Bulletin. 28, 2, 1983, pp. 75.86.

18. Hadagali, P B. Frequently cited periodicals by Indian agricultural economists : A citation analysis. Iaslic Buttletin. 28, 1983, pp. 59-66.

19. Hawkes H E: Literature of geology. Geotimes, 10, 1966, p. 23-24.

20. Henkle H H. Periodical literature of biochemistry. Bull. Medical. Lib. Assoc. 27, 1988, p. 139-147.

21. Humayoon Kabir S. Authorship trend and solo research in bibliometrics : A bibliometrics study. Library Science with a slant to Documentation and Information Studies (Accepted).

22. Humayoon Kabir S. Citations in Indian agricultural research : An analysis. Annals of Library Science & Documentation, 37, 3, 1990, p. 102-109.

23. Humayoon Kabir S. Citations in World bibliometrics research : An analysis. Library Science with a slant to Documentation and Information Studies. 30, 4, 1993, p. 155-159.

24. Karisiddappa C R, *Maheswarappa B S and Shirol M V*: Authorship pattern and collaborative research in psychology Iaslic Bulletin. 35, 2, 1990, p. 73-78.

25. Ken, M C: Bradford's law and the journal titles cited by research scholars in Zoology at Ibadan University Nigeria. Iaslic Bulletin. 34, 3, 1989, p. 97-104.

26. Lawani S M. Bradford's law and the literature of agriculture, International Library Review. 5, 1973, pp. 341-350.

27. Lawani S M and Seriki, T A B. Some characteristics of the World Literature on rice. International Rice Commission Newsletter. 23, 1, 1974, pp. 1-5.

28. Lopez S J. Characteristics of cassava literature. Cassava Newsletter. No. 8, 1980, pp. 10-15.

29. Maheswarappa, B S and Mathias S A: Research collaboration in different disciplines of applied sciences in India (1965-83) : A bibliometric study. Iaslic Bulletin. 32, 1987, pp. 105-114.

30. Maheswarappa B S and Nagappa B: Indian phytopathology literature : A bibliometric study based on review of plant pathology. J. of Library and Information Science. 9, 1, Jun 1984, pp. 36-47.

31. Nag D K, et al: Pattern of use of earth science literature by Indian geologists. A citation analysis. Proc. 15th Iaslic Conf. Bangalore, pp. 101-108.

32. Narendra Kuman P. An international bibliometric survey of periodical articles on bibliogrical control. Mlisc dissertation 1994, Bangalore University, p. 83.

33. Narin F. Evaluative bibliometrics : The use of publication and citation analysis in the valuation of scientific activity. NTIS PB 252339/AS. New Jersey, Computer Horizon Inc., 1976, p. 500.

34. Nijagunappa R and Nijagunappa P. Core journals used by Indian geoscientists (1979-82). J. Geol. Soc. India. 26, 1985, pp. 10-15.

35. Noor Martyvonee M. A quantitative analysis of the research articles published in core library journals of 1980. Library and Information Science Research. 7, 1985, pp. 261-273.

14

INFORMATION USE PATTERN OF AGRICULTURAL ECONOMISTS IN INDIA–A BIBLIOMETRIC STUDY

M CHANDRAKUMARAN NAIR

0. INTRODUCTION

Development of a society depends to a great extent on how best the confronting social, economic, political and cultural problems are understood by the commonman, policy makers, executors and legislators. The ultimate aim of all economic activity is human welfare. The resources of every nation is limited. For bettering the human living condition, the main thrust is that every nation should manage its resources efficiently with a prime objective of alleviating poverty and backwardness. It is in this context the economists are called upon to play a crucial role in the objective and intelligible analysis of various economic problems in their correct perspective and to evolve proper solutions.

Being the fifth factor of production, information has become an essential raw material for the proper analysis of various economic issues which is so diversified in our modern society. Information in the form of data, both raw and processed, is heavily relied upon by the social science information users. It presupposes the need for local, regional, national and international data both published as well as unpublished. Obviously these data materials form the main source of information for such users.

It can be seen that unlike scientists, the social scientists use a variety of information resources such as legislative proceedings, enactments, newspapers, newsletters and bulletins, committee/commission reports, addresses and proceedings of political parties, office files and records, archival materials, time series data materials, etc. An observation is also made that the social scientist generally use retrospective information as well as current information with the same degree of importance. It is in this context, a biotelemetric study is carried out based on the citation pattern of economists working in Centre for

Development Studies (CDS) Trivandrum. As the Centre conducts various studies at Regional and national level, and the study group on which the study was conducted comprise people from many parts of India, the information use pattern of this sample group well exhibit a national pattern.

Citation method is used in this study to understand primarily the information use patterns of the agricultural economists in CDS, one of the leading research Centres in the disciplines relevant to developments in India. The present study takes in to account only the working papers in agricultural economics and its related areas produced by the development economists at the Centre. The research output of the CDS in the form of working papers during 1971-85 is 209. Among them 85(40.6%) are studies in agricultural economics and its allied subjects. The present study covers all the 85 working papers.

In order to get an insight into the specific subject areas in which the studies were concentrated, the working papers were subjected to a primary subject analysis. It is seen that 'Farm production' is the area where the studies were more concentrated. When 10.7% the working papers have been produced in this speciality alone, 'agricultural growth' and livestock economics' ranked next to it with an equal share of 8.33% of the working papers each. Agrarian relations account for 5.95%. Land tenure/ownership, fisheries economics, economics of water management, Farm wages etc. are represented equally with 4.76% each and other specific subjects like 'diary economics' showed lower degree of concentration.

When we examine the area of study of the working papers (Table 1), it is quiet interesting to see that 48.23% of the studies were conducted in the wider national context while 37.65% were devoted to problems specific to Kerala.

Table-1 Area-wise breakup of papers

Area	Number of papers	Percentage	Cumulative percentage
India	41	48.23	48.23
Kerala	32	37.65	85.88
Other states	5	5.88	91.76
International (comparative studies)	5	5.88	97.64
General	2	2.36	00

The foregoing table supply essential evidence to the fact that the Centre has given more emphasis to empirical problems rather than to theoretical ones. This has been a matter of criticism in academic circles which has been pointed out by K.G. Kumar in his article 'Understanding Indian Social Reality' appeared in Business India.[1] It states that CDS has developed an anti-theoretical bias due to this overemphasis to empirical studies to the near total exclusion of major theoretical work, which consequently diverted the Centre's attention from developing a systematic critique of non-classical economics. But a closer look at the objectives at the Centre justifies this bias because, it is stated clearly that in studies on various aspects of development...... both local and national of current interest, the stress in all cases is on empirical studies, informed by a historical perspective and keeping in view the broader social contact of economics problems.[2]

1. ANALYSIS OF CITATIONS

Citations collected from the 85 working papers constituted a heterogenous group of different types of documents such as journal articles, books, research monographs, working papers, serials, reports, conference proceedings, enactments rules-commentaries, etc, official records, newspapers and dissertations.

The citations in the working papers distinctly authenticate the statements given in the text and so their validity as indicators of use is more definite than the references generally given in support of the text as a whole. There are also footnotes used to substantiate ideas, but all of them may not find a place among the references given at the end of the paper. In such cases, footnotes were also counted as separate citations. When the same reference appears in different forms such as working paper, part of a book or journal, etc. it is counted in the specific bibliographic form in which it is cited.

The most cited corporate author was the Directorate of Economics and Statistics of the Ministry of Agriculture, Govt. of India with 35 citations. Most of its publications are serials concerning time series data. A citation referring to various issues of a serial for different years is treated as a single citation unless it is otherwise specified.

The above analysis substantiate well the heavy use of these data sources by the economists.

Rate of self citation

Practice of self citing is common in social science as it is in science. Citations to ones own previous works are known as self citations. But in this study, along with the true self citations, citations to works of colleagues in the CDS were also considered. Total number of self citations counted in this way was 163 (11.53%) and among them only 66 (4.68) were true self citations.

The rate of self citation seems to be high, but the percentage of citations to works of colleagues is more than the percentage of true self citations. This may be possibly related to the fact that the contribution in Agriculture and rural economy made by the core of noted economists in CDS were highly valued in the research activities by the Centre.

Types of documents cited

In the analysis of citations on the basis of its form, it has been found that a variety of document sources are used by the economists in their pursuits. However, only ten categories were found adequate in this study as they were observed as significant types. Thus different sequences were formed for cited journal articles, books, research monographs/working papers, serials, reports, conference proceedings, enactments rules-commentaries, etc.

Table-2 Type of documents cited in working papers

SI. No.	Type of documents	Total No. of citations	Percentage	Cummla-tive %
1	Journal articles	391	27.71	27.71
2	Books	388	27.5	55.21
3	Research monographs/ working papers	240	17.01	72.22
4	Serials	216	15.31	87.53
5	Reports	98	6.95	94.48
6	Conference proceedings	39	2.76	97.24
7	Enactments-rules-commentries	12	0.85	98.09
8	Official records (manuscripts)	12	0.85	98.94
9	Newspapers	10	0.71	99.65
10	Dissertations	5	0.35	
	Total	1411	100	100

Broad Categories

Broad categories were made in order to ascertain the relative use of monographic and periodical literature. Periodical literature includes all those serials having regular and irregular periodicity, journals and news papers. It was upon monographic documents (56.28%) but not to the extent it was observed in the study made by Kaula and Singh.[3] They found that economists in India rely heavily upon monographic information material. This might be related to the fact that the research in CDS is oriented towards practical problems. And it might be probably an additional factor that the pattern of information needs in agricultural economics and use slightly differ from those observed in the general field of social science or a more defined field as economics. In the social science citation index (1975 edition), the percentage of citations to journal and non-journal items was 47 and 53.[4]

Special categories

In the analysis of 10 special categories formed on the basis of type of cited documents as shown in Table 2, journal articles are found to be the most preferred source of information (27.71%) But books too get prominently cited (27.5%). Working papers and other research papers shared 17.01% of the total citations, while serials were represented by 15.31%. Reports of committees and commissions were cited more (6.95%) than conference proceedings (2.76%)

The above findings exhibit a close similarly to that of the INFROSS survey.[5] INEROSS study inferred that for social scientists books and periodicals are more or less equally important in their pursuits.

Fletcher has given a finding in support of the high rate of use of journal literature. In general, social scientists do not mark a very high level of citations of journals, but 'economics' in this respect seems more like science and technology.[6]

Government Publications

Government publications in a separate analysis indicated considerable use comprising 21.47% of citations. This percentage is mainly contributed by the publications of departments of Central and State Governments, Census publications, National sample survey reports, Foreign government publications etc. The statistics shows that for agricultural economists, Govt. publications are accepted sources of useful information.

Chronological Distribution of cited documents and the Obsolescence of Literature

Chronological distribution of cited documents shown in the table 3 reveals that there is a steady increase in the use of publications of 1950 to 1979. But the rate of use show a decreasing tendency for publication from 1980 onwards. It is a striking point that among all types of documents cited, the heaviest use is observed for the literature published during 1970-79. 54.21% of the journal articles, 49.74% of the books, 53.74% of the research papers, 51.85 of serials and 38.76% of the reports cited in the working papers belong to this period.

Publications of the sixties and eighties too have got cited prominently, but the literature emerged in fifties and earlier is far less cited.

Serials published during 1960-69 received remarkable number of citations. Conference proceedings from 1975 are found cited more and this rate of use remains stable till the end of 1981.

The above findings support Garfield's view that for 'literature which is mainly research front', the citation practices in social science may exhibit the character of natural sciences, that is, most of references will be to items published recently. As per age statistics for 1974 social science citation index.[7] the percentage of citations to literature less than five years old is 44.85, while citation to literature more than 15 years old is 21.03. And the states that the documents cannot be put into 'mutually exclusive boxes' labelled with narrow disciplines because of the cross disciplinary trends. The findings from this study also agree to a great extent with Garfield's apprehension about the inferences of INFORS study regarding the considerable difference between the age statistics for citations of Scientists and social scientists. Garfield believes that researchers in social science cite recent items at a rate more than what was revealed by INFOROSS study.

This is further endorsed by a separate analysis of the citations in the working papers belonging to a specific year, ie, 1983 in which CDS produced the largest number of working papers in agriculture and rural economy. This was carried out because, a clear indication as to how far back the researchers go and use the information sources seems to be difficult when the citations cover over a period of 14 years and the results are stated in Table 3. This study gave more dependable

Table-3 Chronological Distribution of Cited Documents

Period	Journal articles (%)	Book (%)	Research mono-graphs/Working	Serials(%)	Reports (%)	Conference Proceedings Total (%)	(%)
1800-99	1 (0.25)	7 (1.8)	-	4 (1.85)	3 (3.06)	-	15 (1.09)
1900-09	-	3 (0.77)	-	1 (0.46)	-	-	4 (0.29)
1910-19	-	2 (0.52)	-	-	-	-	2 (0.01)
1920-29	2(0.51)	2 (0.52)	1 (0.41)	1 (0.46)	1 (1.02)	-	7 (0.5)
1930-39	8 (2.04)	9 (2.30)	-	5 (2.31)	7 (7.14)	-	29 (2.11)
1940-49	7 (1.79)	6 (1.55)	4 (1.67)	1 (0.46)	2 (2.04)	-	20 (1.45)
01950-59	16 (4.09)	21 (5.37)	11 (4.58)	13 (6.01)	17 (17.34)	1 (2.56)	79 (5.76)
1960-64	44 (11.25)	23 (5.92)	12 (5)	27 (12.5)	12 (12.24)	1 (2.56)	119 (8.67)

Period	Journal articles (%)	Book (%)	Research mono-graphs/Working	Serials(%)	Reports (%)	Conference Proceedings Total (%)	(%)
1965-69	39 (9.97)	69 (17.78)	13 (5.42)	31 (14.35)	14 (14.28)	7 (17.95)	173 (12.6)
1970-74	96 (24.55)	100 (25.77)	31 (12.91)	68 (31.48)	20 (20.4)	5 (12.82)	320 (23.32)
1975-79	116 (29.66)	93 (23.97)	98 (40.20.37)	18 (18.18.36)	10 (18.36)	10 (25.64)	379 (27.6)
1980-81	34 (8.69)	30 (7.73)	29 (12.08)	11 (5.09)	1 (1.02)	9 (23.07)	114 (8.30)
1982-83	20 (5.11)	16 (4.12)	22 (9.17)	8 (3.7)	2 (2.04)	4 (10.26)	72 (5.25)
1984-85	8 (2.05)	1 (0.26)	10 (4.17)	2 (0.92)	1 (1.02)	2 (5.13)	24 (1.75)
Total	391	388	240	216	98	39	1372 (100)
		6 unpublished	9 undated				

information about the use of current literature. As different from the social scientists, economists are reported to be relying more on current literature and the present study indicates that the degree of this relevance is even higher in case of agricultural economists.[8]

Table-4 Age of Citations cited in the working papers of 1983

Period	Number of citation	Percentage
Before	1940	4 1.74
1940 - 49	2	0.87
1950 - 59	15	6.55
1960 - 69	30	13.1
1970 - 74	49	21.4
1975 - 79	51	22.27
1980 - 83	73	31.88
Total	229	100

Citation undated

31.88% of the literature used belonged to the current period of 4 years (1980-83). It also needs special mention that the literature of 1970s again figures prominently establishing its relevance by 44.65% of citations. It may be further analysed to indicate the obsolescence ofliterature in agricultural economics. Going backward chronologically, the citation rate diminished. Literature prior to 1950s, shared only minor fractions (below 3%). Therefore, it may be conjectured that the literature earlier to 1950s is the least significant for agricultural economist. Literature of 1950s found a little improved use, but literature of 1960s found moderate use, and literature of 1970s use of high use. Literature of 1980s, although published in a shorter span of 4 years, is the most used.

In short, it may be said that the useful informational material for the agricultural economist is largely confined to the last 20 years.

The highest rate of use of information published in seventies as evidenced in the analysis of citations in working paper seems to be directly related to the availability of data.

Data availability

Over-riding interests of economists on current data got reflected in their use. In relation to the quantity of information generated in the previous years, the decade of seventies must have witnessed an unprecedented advance in the proliferation of current data drawing the immediate attention of agricultural economists. This might be probably related to the following reasons which made possible active research and investigation in the field.

i) Green Revolution :

Intensified research activities were conducted in the agrarian sector in India during this period. Though Green revolution was initiated in 1960s, the effects on a full fledged scale were experienced in 1970s. Naturally this might have led to numerous technoeconomic studies, technical studies and economic studies such as its impact on poverty, income distribution, employment, farm production, productivity etc. Also, studies on influence of irrigation, fertilizer use, high yielding varieties, etc. on growth of output were taken up. It might have further prompted studies on the effect of green revolution on each variety of crop which led to the realization that green revolution was infact 'Wheat Revolution'.

ii) ICSSR :

A number of social science research institutes came up in India during the period since the inception of Indian Council of Social Science Research in the year 1969. As given in the annual report,[9] Promotion of Research is the most important responsibility of the council. This is being pursued through a number of schemes aimed at building up the infrastructure for research, financing of research projects, awarding research fellowships, providing grants to meet the expenditure for assisting training programmes, etc.

iii) Need for agricultural growth and high productivity :

The wide spread belief that the major factor behind the industrial stagnation in the pre-seventies was the low productivity and stagnent agricultural growth. This phenomenon had been rightly pointed out by several economists like S.L. Shetty and K. N. Raj.[10,11]

iv) Land Reforms :

The implementation of land reforms in India in some states in

1970s had resulted in a number of studies dealing with the different aspects of it such as farm size and productivity relationship, mode of production, pattern of ownership, etc.

Country-wise Distribution of cited documents

An analysis of the cited documents by country of origin yielded the data presented in Table 5.

Indian publication are the most cited. 39.69% of the books 54.58% of the research papers, 95.37% of the serials, cent percent of the reports, 51.28% of the conference proceedings, 62.4% of the journal articles cited are of Indian origin.

These findings agree with the conclusions of the INFROSS studies that social scientists are more engaged in research on local emperical problems and this justified their high dependence on Indian publications in this study. Prakash B. Hedagall in a recent citation study on the frequently cited periodicals by Indian agricultural economists has clearly brought to light that they mostly depend on indigenous literature.[12] It warrants the need for organised documentation of Indian Publications.

United States has ranked next to India in the case of periodicals while United Kingdom comes in second position (25.5%) in respect of cited books. Chinese and Japanese documents may be expected to be of high relevance to this study group, but the table reveals the citation rate of books from these countries as 3.09% and 5.92% respectively.

As far as the research monographs are concerned, substantial use of documents published by International Organizations like FAO, ILO, OECD, ARTEP has been observed. As given in the Table-5, 22.91% of the research papers represents intergovernmental organizations.

Language-wise Distribution of the cited documents

English language documents have a dominance over others. 98.72% of the periodicals, 95.88% of the books and 99.58% of research papers were in English. Inspite of the over emphasis to local problems in the research programmes in social sciences, it seems that the language favoured by the specialists for the publication of research findings of English (This has been identified in the earlier studies too)[13] Indian Scientists as well as social scientists do always publish more in English and consequently their dependence of English language documents is more. But the need for Malayalam documents, and translations of foreign language documents in Russian, Japanese, Chinese and French cannot be totally ignored.

Table-6 Language-wise Distribution of cited documents

Types of Documents	Citation in English	Malayalam	Russian	Chinese	Japanese	French
Journal articles	386 (98.72%)	3 (0.77%)	-	1 (0.26%)	1 (0.26%)	-
Books	372 (95.88%)	12 (3.09%)	3 (0.77%)	-	-	1 (0.26%)
Research monographs/ working papers	239 (99.58%)	1 (0. 42%)	-	-	-	-
Serials	216 (100%)	-	-	-		
Reports	98 (100%)	-	-	-	-	-
Conference Proceedings	39 (100%)	-	-	-	-	-
Total	1350 (98.4%)	16 (1.17%)	3 (0.21%)	1 (0.07%)	1 (0.07%)	1(0.07%)

Analysis of Citations to Journal Articles

Detailed analysis were done for the journal citations to have an insight into the use of journals. Rank list of cited periodicals (Table 7) throws much light on various aspects of the same. With the highest citation count sharing 140 citations (42.55%), the economics and political weekly (EPW) has emerged as an indispensable information source for economics research. It is not a periodical special to Agricultural economics. Kaula and Singh also found 'Economic and Political Weekly (EPW) as the title most favoured by economists in India in their study on the characteristics of information material used by Indian Social Scientists.[14] Indian Journal of Agricultural Economics' which carries mostly the current information on agriculture and the related problems finds only a 2nd rank having 39 citations (11.85%). It may be due to the fact that EPW is considered an excellent representation of Indian scholarship in economics and it carries the most nascent research news and scholarly articles. Articles in this journal are not of general nature, but embrace all specialised areas within the broader field of coverage. It is found that the first 31 titles are most productive and 329 citations are concentrated on these 31 titles as shown in the graph. Remaining 62 citations are distributed in 49 titles.

Table-5 Country-wise distribution of cited documents

Country	Journal articels (%)	Books (%)	Research monographs/ working Papers (%)	Serials (%)	Reports (%)	Conference Proceedings	Total (%)
India	244 (62.4)	158 (40.72)	131 (54.58)	206 (95.37)	98 (100)	20 (51.28)	857 (62.69)
United States	80 (20.46)	66 (17.01)	12 (5)	2 (0.93)	-	1 (2.56)	161 (11.77)
United Kingdom	36 (9.2)	102 (26.28)	11 (4.58)	1	-	-	150 (10.97)
International Organizations	-	8 (2.06)	55 (22.91)	4(1.85)	-	9 (23.08)	76 (5.56)
Japan	5 (1.28)	23 (5.92)	9 (3.75)	-	-	6 (15.38)	43 (3.14)
China	2 (0.51)	12 (3.09)	6 (2.5)	2 (0.93)	-	-	22 (1.6)
Other Countries	12 (3.07)	-	2 (0.83)	1 (0.46)	-	3 (7.69)	18 (1.31)
Netherlands	12 (3.07)	4 (1.03)	-	-	-	-	16 (1.17)

Country	Journal articels (%)	Books (%)	Research monographs/ working Papers (%)	Serials (%)	Reports (%)	Conference Proceedings	Total (%)
Russia	-	12 (3.09)	-	-	-	-	12 (0.87)
Norway	-	-	9 (3.75)	-	-	-	9 (0.65)
France	-	3 (0.77)	-	-	-	-	3 (0.21)
Total	391 (100)	388 (100)	240 (100)	216 (100)	98 (100)	39 (100)	1367 (100)
			For 5 citations countries not identified				

Table-7 Ranked list of Cited Pariodicals

Sl. No.	Rank No.	Name of the journal	Country of origin	Year of origin	No. of citations	Cumulative citation	Percentage of citation
1	1	Economic and Political Weekly	India	1949	140	140	42.55
2	2	Indian Journal of Agricultural Economics	India	1946	39	179	11.85
3	3	Indian Economic Review	India	1966	11	190	3.34
4	4	Current Anthropology	USA	1960	9	199	2.74
5	5	Indian Economic and Social History Review	India	1963	8	207	2.43
6	5	Indian Journal of Materology and Geophysics	India	1958	8	215	2.43
7	5	Journal of Political Economy	USA	1892	8	223	2.43
8	5	Journal of Economic History	USA	1941	8	231	2.43
9	6	Social Scientist	India	1972	7	238	2.13
10	6	American Journal of Agricultural Economics	USA	1919	7	245	2.13
11	7	American Economics Review	USA	1911	6	251	1.82
12	7	Indian Journal of Agricultural Science	India	1931	6	257	1.82
13	7	Quarterly Journal of Economics	USA	1887	6	263	1.82
14	7	Review of Economics & Statistics	Netherlands	1919	6	269	1.82
15	8	Economic Development and Cultural Change	USA	1952	5	274	1.52

Sl. No.	Rank No.	Name of the journal	Country of origin	Year of origin	No. of citations	Cumulative citation	Percentage of citation
16	8	Journal of Indian Society of Agricultural Statistics	India	1949	5	279	1.52
17	8	Economics Journal	UK	1891	5	284	1.52
18	8	Sankhya	India	1934	5	289	1.52
19	9	Journal of Development Studies	Netherlands 1964	4	293	1.22	
20	10	Cambridge Journal of Economics	UK	1977	3	296	0.91
21	10	Demography	India	1972	3	299	0.91
22	10	Developing Economics	Japan	1963	3	302	0.91
23	10	Enquiry	UK	1970	3	305	0.91
24	10	Frontier	India	1968	3	308	0.91
25	10	Indian Historical Review	India	1974	3	311	0.91
26	10	Journal of Peasant Studies	UK	1973	3	314	0.91
27	10	Journal of the Royal Statistical Society Series A	UK	1839	3	317	0.91
28	10	Manchester School of Economics & Social Studies	UK	1930	3	320	0.91
29	10	New Left Review	UK	1981	3	323	0.91
30	10	Past and Present	UK	1952	3	326	0.91
31	10	Water Resources Research	USA	1965	3	329	0.91

Distribution of Journal Citations

The relationship between the number of citations and its distribution has received considerable attention. However, the graph shows that the journal literature in agricultural economics does not follows strictly the Bradford's law of scattering as we find in applied science in general and pure sciences in particular.

Out of the 80 journals referred, 31 were cited not less than 3 times as given in the rank list. If titles were cited twice and the rest (39) journals were used only once. It can be seen that they are important journals in the field of economics but are of interdisciplinary nature and thus required limited use in this area selected for the study. These journals might also be of interest study. Journals, 'Indian Historical Review and Demography India' are ranked in the 10th position having only 3 citations each. But the same journals like current anthropology', social scientist, 'Indian Economics and Social History Review', 'Economic Development and Cultural changes', Social action etc. are primarily of interest to scholars in social science in general but they have also been substantially used by agricultural economists.

31 titles listed in the table have been frequently cited by the users. The ranked list of the journals here shows (Table 7) a close similarity to the ranked list of the journals in a more or less similar study on

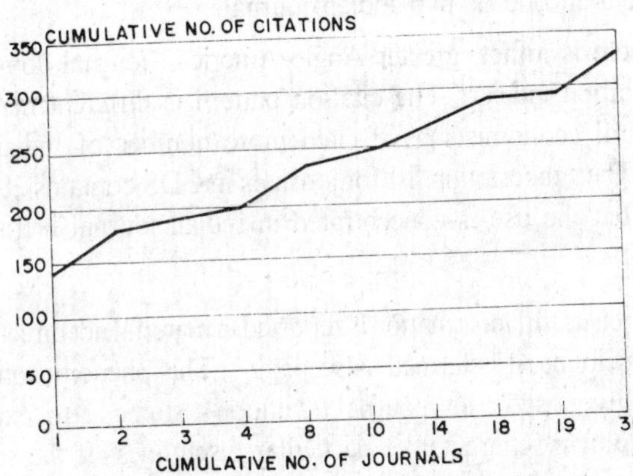

Fig. 1 Distribution of Journal Citations

economists by A. L. Kapoor.[15] The journals 'Economic and Political weekly', 'Indian Journal of Agricultural Economics, Journal of Political Economy', American Journal of Agricultural Economics, Indian Economic Review etc. which come in the first 6 ranks occupied more or less a similar position within 6 ranks in the study conducted by Kapoor.

It may be stated that use of some titles cannot be confined to particular subject, since almost all the journals are found used by the Development Economist irrespective of their area of specialisation. One reason for this may be the interdisciplinary nature of these journals as stated earlier.

Indian and Foreign Journals

Analysis of the journal citations were done in two ways. Firstly, the total number of citations were grouped under the respective country of origin of the journals. It showed that 244 citations (62.44%) were from Indian journals. 20.46% and 9.2% of the journals citations appeared in the journals published from USA and UK respectively.

Secondly, the titles of the journals were grouped country wise irrespective of the number of citations. It revealed a different picture, ie, of the 80 titles used, only 30 titles (37.5%) were published from India. But 26 titles (32.5%) published form USA, 17 titles (21.25%) form UK, and 7 titles published form other foreign countries are also used. The difference in the two analysis is due to the high concentration of citations in one or two Indian journals.

There is infact, greater Anglo-American journal dominance in social citation index.[16] The citation pattern is different here, as our agricultural economists have cited more number of Indian journal articles. The subscription list for journals in CDS contains 296 foreign journal, but the use is concentrated in Indian journal literature.

Seepage

When useful information is reported in material not directly related to that discipline, it is termed as seepage.[17] This phenomenon is found increasingly used in journal citation studies to explain the interdisciplinary character of particular disciplines. If the seepage is dominant, it will have direct implication to information activities.

It is found that agricultural economists are intense users of general economics journals and social science journals. The use of general economics journals, as indicated by their citation practice, is more than 1 1/2 times of the use of journal literature in agricultural economics, the core area. Specialised journals in the various branches of social science are more utilised than the general social science journals. 8.75% of the citations some from journals in Development Studies.

It has been observed in earlier studies that agricultural economists publish extensively in documents of the fields of poultry science, dairy science, food science etc. But it is worth noting in the present study that 5.63% of the articles cited belong to the science journals covering meterorology and geophysics, atomic science, dairy science, etc. Thus it reveals that the phenomenon of seepage is not very high. Agricultural economists here rely on economics, social science and development studies for most of their journals literature. It leads to the inference that, inspite of the inter-disciplinary nature of the field of agricultural economics, its journal literature does not cut across a wide spectrum of disciplines not directly related to it.

Dispersion

Table-8 Dispersion of Journal citations by disciplines of source journal

Subject	No. of Journals (%)	No. of Citations	Average (%)
1. Agricultural Economics	13 (16.25)	67 (17.17)	5.15
2. Economics (G)*	18 (22.5)	186 (47.65)	10.33
3. Economics (S)*	10 (12.5)	33 (8.48)	3.3
4. Social Science (G)	10 (12.5)	19 (4.87)	1.9
5. Social Science (S)	13 (16.25)	45 (11.53)	3.46
6. Development Studies	7 (8.75)	18 (4.6)	2.57
7. Science (S)	8 (10.2)	22 (5.64)	2.75
Total	78	390	

*G = General *S = Specialised (Branches of the host subject)

Analysis of the dispersion of periodical literature is also beneficial in the sense that is may provide certain guidelines for its acquisition. In this contact, it is useful to find out the extent of scattering or dispersion of useful information within each of these areas in different periodicals. General economics periodicals show the least dispersion or in other words the highest concentration by supplying an average of 10.33 journal citations. Dispersion of useful information is found to be higher in the area of agricultural economics. Next to this, in the increasing order of dispersion are rated the journals in Social science, Economics, Science and Development studies.

2. CONCLUSION

Citation method is found increasingly applied in studying information use pattern of development scientists and the findings of this study also carry wide implications. However, the results reveal with two reservations. Firstly, as the citing documents spread over from 1972-1985, current information needs and use patterns cannot be specially projected.

Secondly, citation analysis may reveal information needs and use patterns as they stand expressed by actual use. Actual use is strongly influenced by the availability of information sources and bibliographic services for the user. Hence, it may at best, given an objective, but partial picture of the problem. The above limitations are to be resolved by adopting additional techniques to supplement the findings. Nevertheless, the results of this investigation carry wider implications to the overall design and improvement in quality and economy of library and information services.

REFERENCES :

1. Kumar, K. G: Understanding Indian social reality, Business India. No. 142, Aug. 1983: A73-77
2. Centre for Development Studies (Trivandrum) Annual Report, 1978-79 Trivandrum: CDS. p6.
3. Kaula, P. N. and Singh, S. N: Characteristics of information material: A bibliographic Investigation in the field of social sciences in India. Herald of Library Science 19(4) 1980: p.290.
4. On the literature of the social sciences and usage and effectiveness of the Social Sciences Citation Index. In Garfield, Essays of an information scientist, Philadelphia: ISI, 1977. p.551.
5. Bath University: Investigation into information requirements of social scientists. Vol. 1: Text. Cleverdon: Bath University, 1970.

6. Fletcher, John: A view of the literature of economics. Journal of Documentation 28(4) 1972: p.286.
7. On the literature of the social science, Op. Cit.p.551.
8. Fletcher, John. Op.cit. p. 283.
9. Indian Council of Social Science Research, Annual Report 1982-83.
10. Shetty, S. L: Structural retrogression in the Indian economy since mid sixties. Economics and Political Weekly 13 (6 & 7) 1978: p185-244.
11. Raj, K. N: Growth and stagnation in Indian industrial development Economic and political weekly 11 (5, 6 & 7) 1976 : pp. 223-236.
12. Hadagali, Prakash, B. Frequently cited periodicals by Indian agricultural economists. IAIALIC Bulletin 29(2) 1983: p 63.
13. Earle Penelope and Vikery Brain. Social science literature use in the U. K. as indicated by citations. Journal of Documentation 25(2) 1969: p 130.
14. Kaula and Singh. Op.cit. p. 551
15. Kapoor, A. L. Use pattern of periodicals in economics : A case study of Ratan Tata Library. Library Herald 23(2 & 3) 1984: pp. 82-83.
16. On the literature op. cit. p.551
17. Kaula and Singh, op. cit. p.
18. Littleton, Isaac T. The literature of agricultural economics: Its bibliographic organization and use, Library Quraterly 39(1) January 1969: p. 143.

15

INFORMATION NEEDS, USE PATTERN AND USE BEHAVIOUR OF SOCIAL SCIENTISTS : ANALYTICAL APPROACH THROUGH CITATION METHOD

JESSY JOHN

0. INTRODUCTION

Citation analysis is one of the popular methods employed in recent days for the identification of core documents in various subject fields or for a particular scientific community in a geographical proximity. In fact, citation study is one of the effective and indirect methods to understand the information requirements of users. Citation method is used in the present study to understand the information needs, use pattern and use behaviour of social scientists. The main objectives of this study are to identify:

(i) Average number of citations;

(ii) Authorship pattern

(iii) Distribution of citation by different forms of sources;

(iv) Agewise distribution of citations;

(v) Distribution by place of publication etc.

1. ANALYSIS

The Ph.D theses in social science accepted by the Kerala University, Thiruvananthapuram during 1980-1991 are taken as the source data for the present study. 'Citation approach' is the methodology choosen for this study which is based on the analysis of the bibliographic references appended to the theses. For this study, 37 Ph.D theses were selected from various disciplines in social science. Details regarding number of theses taken from various disciplines are given in table number 1 below.

6260 citations were collected from 37 theses. These citations stand as the baseline data for the citation analysis and interpretation.

Table-1 Break-up of theses in various disciplines in social science

Subject	Year	No. of theses selected
Psychology	1985-90	5
Education	1985-90	5
Geography	1980-90	4
History	1980-90	5
Political Science	1980-90	5
Economics	1985-91	5
Sociology	1985-90	5
Law	1980-90	3
Total		37

1.1. Average of citations

The following table shows the average citations per thesis in various disciplines in social science.

It is observed that thesis of Education with average citations of 242, scores over other subjects. The theses on Psychology secure second position in the average citations where as the theses on Political Science are with the least number of citations.

Table-2 Range of citations in social science theses

Subjects	No. of theses	Total citation	Average
Psychology	5	1069	214
Education	5	1213	242
Geography	4	490	122
History	5	784	157
Political Science	5	586	117
Economics	5	724	145
Sociology	5	788	157
Law	3	606	202

1.2 Age-wise distribution of citations

Chronological distribution of citations reveal the relative use of different publications. Here the analysis shows that publications of 1960s and 1970s have got cited prominently.

1.21 Age-wise distribution of citations in Psychology

Table-3 Age-wise distribution of citations in Psychology

Year of publication	No. of citations	Percentage
Before - 1950	178	16.65
1950 - 1959	232	21.70
1960 - 1969	325	30.40
1970 - 1979	275	25.73
1980 - 1989	59	5.52

The data given in Table 3 reveals that the maximum 'literature cited falls in the decade 1960's with 30.40% and the literature emerged during 1980-89 are less cited.

1.3 Ranking of journals in Social Sciences

Journal citation studies have been used as a means to measure the utility of Journals in general as well as specific areas of social science. Ranking of Journals in each discipline in social sciences is given below.

Table-4 Cited periodicals in Psychology

Name of the Journal	No. of Citation	Rank
Journal of Counselling Psychology	50	I
Journal of applied Psychology	16	II
Journal of Social Psychology	15	III
Exceptional children	9	IV
Personal guidance Journal	7	V
Journal of educational psychology	6	VI
British Journal of educational psychology	5	VII
Human relations	4	VIII
American political science review	4	VIII

Occupational psychology	4	VIII
Vocational guidance quarterly	4	VIII
Journal of abnormal psychology	3	IX
American Journal of sociology	3	IX
Indian Journal of criminology	3	IX
Personal psychology	3	IX
Journal of psychological research	3	IX
Tamil Nadu Police Journal	3	IX
British Journal of Psychology	2	X
Journal of Kerala Studies	2	X
Journal of Criminal Law	2	X
Journal of statistical society of London	2	X
British Journal of Social and Criminal Psychology	2	X
Journal of Psychology	2	X
Journal of educational research	2	X
Journal of marriage and family	2	X
Indian Journal of social work	1	XI
Psychological bulletin	1	XI
Quarterly journal of child behaviour	1	XI
Journal of Health and Social behaviour	1	XI
Journal of Consulting psychology	1	XI
Journal of social science medicine	1	XI
Educational Review	1	XI
American psychologist	1	XI
American Journal of orthopsychiatry	1	XI
Clinical paediatrics	1	XI
Journal of the institute of educational Research	1	XI
Psychiatry	1	XI

Mental Hvgene	1	XI
Sociological bulletin	1	XI
Journal of American Sociological Association	1	XI
Crime and society	1	XI
Social welfare	1	XI
American Journal of psychiatry	1	XI
Indian police journal	1	XI
British Journal of educational psychology	1	XI
Public opinion quarterly	1	XI
American Sociological Review	1	XI
Journal of genetic psychology	1	XI
Social forces	1	XI
American political science review	1	XI
Psychological reports	1	XI
Journal of social issues	1	XI
Review of educational research	1	XI
Annual review of psychology	1	XI
Journal of personality	1	XI
Journal of vocational behaviour	1	XI
The American College	1	XI

Data given in table number 4 clearly show that the 'Journal of Counselling Psychology' is the most productive journal in respect of citations. 'Journal of applied psychology' ranks second whereas 'Journal of social psychology' goes to third position.

1.4 Distribution of citations among different sources

This study helps to understand the information needs, use pattern of various kinds of documentary sources in social sciences among social scientists.

1.41 Type of documents cited in Psychology

From the analysis, it can be observed that periodical is the most cited source (50.80%). The next position goes to books (41.44%).

Table-5 Type of documents cited in Psychology

Type of sources	Total citations	Percentage
Books	443	41.44
Periodicals	543	50.80
Govt. publications	17	1.59
Reports	13	1.22
Conference and seminar Proceedings	23	1.68
Theses	18	2.15
Newspapers	12	1.12

1.5 Authorship Pattern

With a view to identifying the extent of individual and collaborative research the number of authors named in the author statement are counted and analysed. The following table gives the distribution of articles on the basis of number of authors.

1.51 Authorwise distribution of citations in Psychology

It is clear from the table that in Psychology largest number of contributions have been made by single authors (70.63%).

Table-6 Authorship pattern in Psychology

Type of author	No. of citations	Percentage
Single authors	736	70.63
Two authors	224	21.50
More than two authors	82	7.87

1.6 Country-wise distribution of cited documents

An analysis of the countrywise distribution in cited documents obtained the following observations as given below:

1.61 Country-wise distribution of citations in Psychology

The data given in table 7 shows that the United States publications are the most cited (48.19%) UK has ranked next to United States (22.06%). Indian publication has only third position.

Table-7 Country-wise distribution of citations in Psychology

Name of the country	No. of citations	Percentage
India	104	9.64
USA	520	48.19
UK	238	22.06
USSR	12	1.11
Others	205	19.00

2. AGE-WISE DISTRIBUTION OF CITATIONS IN EDUCATION

Age-wise distribution of citation in Education is given in Table 8.

Table-8 Age-wise distribution of citations in Education

Year of publication	No. of citations	Percentage
Before 1950	114	9.40
1950 - 1959	217	17.89
1960 - 1969	419	34.54
1970 - 1979	270	22.26
1980 - 1989	193	15.91

The analysis of citations in Education reveals that the maximum citations fall in the decade 1960's with 34.54%, Citations before 1950 are the least.

2.1 Ranking of periodicals in Education

Table-9 cited periodicals in Education

Name of periodical	No. of citations	Rank No.
Journal of Educational Research	12	I
Journal of Educational Psychology	10	II
Child Development	6	III
Journal of education Research and Extension	6	III

Journal of Psychology	5	IV
Journal of consulting psychology	5	IV
Journal of social psychology	4	V
British journal of psychology	4	V
Journal of experimental education	4	V
Journal of abnormal and social psychology	4	V
Personal and Guidance Journal	4	V
Sociometry	2	VI
NCERT News letter	2	VI
School review	2	VI
Indian psychological review	2	VI
Journal of psychology researches	2	VI
American Journal of psychology	2	VI
Journal of experimental social psychology	2	VI
Journal of applied psychology	2	VI
The progress of education	2	VI
American vocational journal	1	VII
American school board journal	1	VII
Quest in Education	1	VII
Antioch College bulletin	1	VII
Journal of business education	1	VII
Vidyalay Journal	1	VII
Journal of secondary education	1	VII
Bulletin of the international bureau of education	1	VII
American Education	1	VII
Journal of educational measurement	1	VII
Proceedings of science congress	1	VII
School Education	1	VII
Pensylvania school	1	VII
Psychological abstracts	1	VII

Journal of experimental psychology	1	VII
British Journal of medical psychology	1	VII
Scientific American	1	VII
Educational digest	1	VII
California Journal of educational Research	1	VII
Journal of English language teaching	1	VII
American educational research	1	VII
Journal of clinical psychology	1	VII
Adult psychology	1	VII
Journal of genetic psychology	1	VII
Journal of behavioural sciences	1	VII
Indian educational review	1	VII
Journal of social issue	1	VII
Educational review	1	VII
Indian Journal of psychology	1	VII
Journal of psychological research	1	VII
Journal of educational sociology	1	VII
British Journal of educational psychology	1	VII
Journal of marriage and family	1	VII

Out of the 53 journals referred by the Ph.D scholars, 6 journals were referred more than six times. The table shows that the Journal of educational research is the top ranking journal. 'Journal of Educational psychology' ranks second. About 33 journals were cited only once.

2.2 *Types of documents cited in Education*

Table-10 Type of documents cited in Education

Form of source	No. of citations	Percentage
Books	495	40.81
Periodicals	497	40.97
Govt. publications	45	3.71
Reports	43	3.55

Conference and Seminars		1.07
Theses	120	9.89
News papers	--	--

The above table shows that high percentage of citations to books and periodicals where compared with conference proceedings, theses etc. This indicates that books and periodicals are the most preferred channels of information among research scholars in Education.

2.3 Author-wise distribution citations in Education

Table-11 Authorship pattern in Education

Type of authorship	No. of citations	Percentage
Single authors	806	73.40
Two authors	209	19.05
Morethan two authors	83	7.55

The data given in the table above show that the largest contribution has been made by single authors (73.40%) in Education.

2.4 Country-wise distribution of citations in Education

In Education also the publications from USA are the most cited (38.50%). The second rank goes to the publications from India (23.41%).

Table-12 Country-wise distribution of Citations in Education

Name of the country	Name of citations	Percentage
India	284	23.41
USA	467	38.50
UK	280	23.08
USSR	22	1.81
Others	160	13.20

3. AGE-WISE DISTRIBUTION OF CITATIONS IN GEOGRAPHY

Table-13 Age-wise distribution of citations in Geography

Years of publications	No. of citations	Percentage
Before 1950	32	6.53
1950 - 1959	77	15.71
1960 - 1969	139	28.37
1970 - 1979	185	37.76
1980 - 1989	57	11.63

From the above table it can be inferred that the maximum literature cited falls in the decade 1970's with 37.76%.

3.1 Ranking of periodicals in Geography

Table-14 Ranking of cited periodicals in Geography

Name of the Journal	No. of citations	Rank No.
Deep Sea Res.	16	I
Indian J. of Marine Science	14	II
J. Mar. Res	14	II
Trans. Ameer Geophysics	11	III
Limnbol Oceanography	10	IV
Bull. Nat. Inst. Science. India	10	IV
Marine Biol. Assn. India	7	V
J. Geology	7	V
Proc. Indian Academy of Sciences	6	VI
J. Indian Geophys. Union	4	VII
J. Geophysics Res.	4	VII
Mar. Geology	4	VII
Okeanolgia	4	VII
J. of Geophysics	4	VII
J. Scd. Petrol	3	VIII

Aust. J. Mar Fresh Water	3	VIII
Indian Ocean Exped. News letter	3	VIII
Nature	3	VIII
Science	3	VIII
Indian J. Met Geophysics	2	IX
Personal Guidance Journal	2	IX
J. of Counselling psychology	2	IX
Annal Chem. Acta	2	IX
Proc. Roy Soc. ser	2	IX
Australian Geographer	2	IX
J. of Fluid machines	2	IX
The Port Engineer	2	IX
Proc. Amer Society Civil Engineers.	2	IX
Trans. Amer. Society Civil Engineers	2	IX
Intl. Ind. Ocean expedition News letter	2	IX
Ind. J. Fish	2	IX
Atmos and Ocean Phys.	2	IX
Bull. Bingham oceanog	1	X
Estuarine coast Mar. Sc.	1	X
J. Fish Canada	1	X
Amer J. of Sociology	1	X
Vocational guidance quarterly	1	X
J. of occupational psychology	1	X
J. of personality and social psychology	1	X
Educational Research	1	X
The sea	1	X
Indian J. of Animal Science	1	X
Science Rep	1	X
Kiel, Meiresforsch	1	X
Okeanol Akad nauk	1	X

Proc. Indi. Natl Sc. Academy	1	X
Bull National Oceanography	1	X
Curr. Sco.	1	X
Geographical studies	1	X
Ame. Assoc. of pet. Geol Bull	1	X
Pond. J. Mar Res.	1	X
Mahasagar	1	X
J. Ame. Assoc. pet	1	X
Curr. Soil	1	X
J. of Physical oceanography	1	X
Beach Erosion Board Bull	1	X
J. Inst. Civ. Engineers	1	X
Bull. Dept. of Marine Science (Cochin)	1	X
Review of Geophysics	1	X
Tran. Insti. Brit. Geogra.	1	X
Ann. N. Y. Acad. Sc	1	X

The above table clearly shows that 'Deep sea research' is the top ranking Journal. 'Indian Journal of Marine Science and 'Journal of Marine research' ranked second and Transactions of American Geophysics Union' came in the third position. About 14 journals are cited more than four times whereas 28 journals cited only once.

3.2 Type of documents cited by Researchers in Geography

Table-15 Type of documents cited in Geography

Type of Source	No. of citations	%
Books	124	25.31
Periodicals	282	57.55
Govt. Publications	18	3.67
Reports	28	5.71
Conference and Seminars	26	5.31
Theses	12	2.45
News Papers	- -	- -

The above table reveals that the scholars in Geography use periodicals more than any of the sources. It is seen that periodicals are the major source of citation (57.55%) the second preference goes to books (25.31%). The Theses are the least cited source.

3.3 Author-wise distribution of citations in Geography

Table-16 Authorship pattern in Geography

Type of authorship	No. of citations	Percentage
Single authors	227	60.37
Two authors	103	27.39
More than two authors	46	12.24

In Geography also single authors are the most contributed (60.37%) group.

3.4 Country-wise distribution of citations in Geography

Table-17 Country-wise distribution of citations in Geography

Name of the Country	No. of Citations	Percentage
India	87	17.76
USA	213	43.47
UK	82	16.74
USSR	8	1.63
Others	100	20.40

In Geography, the data reveals that United States publications are the main source of information being used in research to the extent of 43.47% publications and India ranks second position.

4. AGE-WISE DISTRIBUTION OF CITATION IN HISTORY

A statement of age-wise distribution of citations in History is given in table No. 18 below.

Table-18 Age-wise distribution of citations in History

Year of publication	No. of citations	Percentage
Before 1950	303	38.65
1950 - 1959	57	7.27

1960 - 1969	108	13.78
1970 - 1979	210	26.79
1980 - 1989	106	13.51

The above table clearly states that researchers in History cited the maximum (38.65%) literature published before 1950. This indicates that old publications are so important among historians for their investigations.

4.1 Ranking of cited periodicals in History

Table-19 Ranking of cited periodicals in History

Name of periodicals	No. of time cited	Rank
Journal of Indian History	14	I
Journal of Kerala Studies	12	II
Indian Historial Review	12	II
Indian Economic and Social History Review	9	III
Kerala Archives News letters	4	IV
Proceedings of Kerala legislative Assembly	3	V
Frontline (Mids)	2	VI
Bhavans Journal (Bombay)	2	VI
Caravan (New Delhi)	2	VI
Illustrated Weekly	2	VI
India Today	1	VII
Link	1	VII
The Week	1	VII

In History, the analysis shows that the use of Journals are less; compared to other subjects. The above table shows that 'Journal of Indian History' is the most cited journal 'Journal of Kerala studies' and 'Indian Historical review' ranked second.

4.2 Forms of documents cited by History

Table-20 Forms of documents cited by Historians

Form of Source	No. of citations	%
Books	5.60	71.43
Periodicals	75	9.57
Govt. Publication	79	10.08
Reports	20	2.55
Conference and Seminar Procee.	16	2.04
Thesis	6	0.77
News papers	28	3.56

In the case of History the highest percentage of cited document is of books ie 71.43% followed by Government publication (10.08%), showing substantial use of Government publications to carry out area studies.

4.3 Authorwise distribution of citations in History

Table-21 Authorship pattern in History

Type of Authorship	No. of citations	%
Single authors	520	92.86
Two authors	34	6.07
More than two authors	6	1.07

The citation in History shows that the larger contributions has been made by the single authors (92.86%).

4.31 Country-wise distribution of citation in History

Table-22 Country-wise distribution of citations in History

Name of the country	No. of citations	%
India	574	73.21
USA	120	15.31
UK	40	5.10
USSR	10	1.28
Others	40	5.10

In History, Indian Publications are the main source of information being used in research to the extent of 73.21% followed by USA and

UK with 15.31% and 5.10% respectively.

5. AGE-WISE DISTRIBUTION OF CITATIONS IN POLITICAL SCIENCE

Table-23 Age-wise distribution of citations in Political Science

Year of publication	No. of citations	%
Before 1950	46	7.85
1950 - 1959	87	14.85
1960 - 1969	78	13.31
1970 - 1979	309	52.73
1980 - 1989	66	11.26

The above analysis reveals that the maximum literature cited falls in the decade of 1970's.

5.1 Ranking of cited periodicals in Political Science

Table-24 Ranking of periodicals in Political Science

Name of periodicals	No. of citations	Rank
Indian Journal of Public Administration	14	I
Economic and political weekly	13	II
Political Science Review	11	III
Economic Weekly	7	IV
Journal of Kerala studies	5	V
Indian Journal of Political Science	4	VI
Chandrika	3	VII
Political quarterly	2	VIII
AICC Economic Review	2	VIII
Public personnel management	2	VIII
Civil service	2	VIII
Administration	1	IX
American political Science Review	1	IX
Focus	1	IX

The table shows that "Indian Journal of public Administration is the most cited Journal in political science. 'Economic and political weekly'; ranked second and political science review' ranked third position.

5.2 *Type of documents cited by researchers in Political Science*

Table-25 Type of documents cited by researchers in political science

Type of Source	No. of citations	Percentage
Books	365	62.29
Periodicals	93	15.86
Govt. Publications	59	10.07
Reports	32	5.46
Conference and seminars	8	1.37
Theses	11	1.88
News papers	18	3.07

From the above analysis, it can be observed that the scholars in Political Science use books (62.29%) more, than other types of sources.

5.3 *Authorwise Distribution of citations in Political Science*

In political science the analysis reveals that single authors are the largest contributors (85.99%).

Table-26 Authorship pattern in Political Science

Type of authorship	No. of citations	Percentage
Single authors	399	85.99
Two authors	56	12.07
More than two authors	9	1.94

5.4 *Country-wise Distribution of citations in Political Science*

The Table 27 shows the overall distribution of citations in political science by place of publication

Table-27 Countrywise distribution of citations in Political Science

Name of the Country	No. of citations	Percentage
India	450	76.29
USA	20	3.41
UK	68	11.60
USSR	18	3.07
Others	30	5.12

The above table shows that in political Science Indian publications are the main sources of information with 76.29% where as publications from UK rank second position.

6. AGE-WISE DISTRIBUTION IN ECONOMICS

Table-28 Age-wise distribution of citations in Economics

Year of Publication	No. of citations	Percentage
Before 1950	42	5.80
1950 - 1959	48	6.63
1960 - 1969	138	19.06
1970 - 1979	259	35.77
1980 - 1989	237	32.74

It is observed from the table that maximum number of literature cited by Economists falls in the decade 1970's with 35.77%.

6.1 Ranking of cited periodicals in Economics

Table-29 Ranking of cited periodicals in Economics

Name of the Journal	No. of citations	Rank
American Economic Review	18	I
Indian Labour Journal	10	II
Economic and political weekly	10	II
Indian Journal of Labour Economics	8	III
Economic Times	7	IV
Journal of political Economy	6	V

Economic Journal	6	V
World Development	5	VI
Cashew Bulletin	3	VII
Indian Economic Journal	3	VII
Engineering Economists	3	VII
The Malayalam Economic Review	2	VIII
Quarterly Journal of Economics	2	VIII
Review of Economic Studies	2	VIII
British Journal of Industrial Relations	2	VIII
Econometrica	1	IX
Quarterly Journal of Economics	1	IX
Journal of American statistical Association	1	IX
Indian Economic Review	1	IX
Applied economics	1	IX
Journal of Econometrics	1	IX
Survey of current Business	1	IX
Journal of income and wealth	1	IX
Journal of Economics	1	IX
Journal of development studies	1	IX
Journal of industry and trade	1	IX
Indian Journal of economics	1	IX
Journal of Public Economics	1	IX
Economic Review	1	IX
Commerce	1	IX
Canadian Journal of Economics	1	IX
Journal of development Economics	1	IX
Demography	1	IX
Chartered secretary	1	IX
Oxford economic papers	1	IX

African Journal	1	IX
Journal of Industrial Economics	1	IX
Journal of finance	1	IX
Journal of financial and structural analysis	1	IX
Artha Vinjana	1	IX
Journal of Business	1	IX .

'American Economic Review' is the most cited and top ranking Journal among the Economists. 'Indian Labour Journal' and 'Economic and Political Weekly' ranked second position' 'Indian Journal of Labour Economics' goes to third position.

6.2 *Type of Documents cites in Economics*

The analysis reveals that the scholars in Economics cited more books (58.70%) compared with other sources. This indicates that books are major channels of information for their research. Their second preferred source is periodicals (26.52%).

Table-30 Type of documents cited by Economists

Type of sources	No. of citations	Percentage
Books	425	58.70
Periodicals	192	26.52
Govt. publications	43	5.93
Reports	41	5.66
Conference and seminars Procee.	7	0.97
Theses	10	1.38
News papers	6	0.84

6.3 *Authorwise distribution of citations in Economics*

In economics, the data shows that single authors are the largest contributors (82.41%).

Table-31 Authorship pattern in Economics

Type of authorship	No. of citations	Percentage
Single authors	478	82.41
Two authors	88	15.17
More than two authors	14	2.42

6.4 Country-wise distribution of citations in Economics

In Economics Indian publications are the main source of information being used in research to the extent of 58.98% followed by USA and UK with 18.24% and 5.80 (Table 32).

Table-32 Country-wise distribution of citations in Economics

Name of the country	No. of citations	Percentage
India	427	58.98
USA	132	18.24
UK	42	5.80
USSR	8	1.10
Others	115	15.88

7. AGE-WISE DISTRIBUTION IN SOCIOLOGY

Table-33 Age-wise distribution of citations in Sociology

Year of publication	No. of citation	Percentage
Before 1950	109	13.83
1950 - 1959	85	10.78
1960 - 1969	162	20.56
1970 - 1979	234	29.70
1980 - 1989	198	25.13

The table makes clear that maximum literature cited by sociologists falls in the decade 1970's with 29.70%.

7.1 Cited periodicals in sociology

Table-34 Ranking of periodicals in sociology

Name of periodicals	No. of citations	Rank
Social welfare	15	I
Journal of social issues	12	II
Population studies	11	III
Social change	8	IV
Economic and political weekly	7	V
American anthropologist	7	V

Man in India	6	VI
Studies in Family Planning	5	VII
Demography	5	VII
Journal of rehabilitation in Asia	4	VIII
Journal of rehabilitation	3	IX
Population and development Review	3	IX
Personality and social psychology	2	X
Journal of Asiatic society of Bengal	2	X
American Journal of sociology	2	X
Demography India	2	X
Social problem	2	X
Social science	1	XI
Journal of Kerala studies	1	XI
Rehabilitation psychology	1	XI
Survey	1	XI
Rehabilitation research and practice review	1	XI
Social work forum	1	XI
Indian Journal of Industrial Relation	1	XI
Indian Journal of social work	1	XI
Kurukshethra	1	XI
Research series	1	XI
Population Bulletin	1	XI
Population	1	XI
Journal of sociology	1	XI
Journal of biosocial services	1	XI
Statistical review	1	XI
Social Biology	1	XI
Journal of social research	1	XI
Proceedings of the Indian Academy of science	1	XI

Journal of Indian Anthrolopological society	1	XI
Anthropological survey of India	1	XI
Eastern Anthropolgist	1	XI
The Anthropologist	1	XI

In sociology 'social welfare' is the most cited Journal. 'Journal of social issues' ranks second and 'population studies' in the third position.

7.2 *Type of documents cited in Sociology*

Table-35 Type of documents cited by sociologists

Type of source	No. of citations	Percentage
Books	436	55.33
Periodicals	240	30.46
Govt. Publications	44	5.58
Reports	31	3.93
Conference and seminar Procee.	23	2.92
Theses	6	0.76
News papers	8	1.02

The above analysis reveals that the sociologists refer books (55.33%) as the main source of information for their research. Their second choice is periodicals (30.46%).

7.3 *Authorwise distribution of citations in sociology*

Table-36 Authorship pattern in sociology

Type of authorship	No. of citations	Percentage
Single authors	580	86.44
Two authors	73	10.88
More than two authors	18	2.68

In sociology the data given in table 21 show that single authors are the largest contributors (86.44%) whereas multiple authorship pattern is relatively very low.

7.4 Country-wise distribution of citations in sociology

In sociology the data makes clear that Indian publications are the most cited (41.37%) followed by USA and UK (24.62% and 11.93%) respectively.

Table-37 Country-wise distribution of citations in sociology

Name of the country	No. of citations	Percentage
India	332	42.13
USA	194	24.62
UK	94	11.93
USSR	6	0.76
Others	162	20.56

8. AGE-WISE DISTRIBUTION OF CITATIONS IN LAW

The analysis (Table 37) reveals that maximum literature cited by lawyers falls in the decade 1970's with 33%.

Table-38 Age-wise distribution of citations in Law

Year of publication	No. of citations	Percentage
Before 1950	155	25.88
1950 - 1959	51	8.42
1960 - 1969	119	19.64
1970 - 1979	200	33.00
1980 - 1989	81	13.36

8.1 Cited periodicals in Law

Table-39 Ranking of cited periodicals in Law

Name of periodicals	No. of citations	Rank
All India Reporter	18	I
Indian Journal of international Laws	17	II
AC L. Review	13	III
Harv. L. Review	10	IV
Law and Soc. Rev.	8	V
Minestoa Law Rev.	6	VI
The police chief	6	VI

Israel Law Review	4	VII
Criminal Justice Review	4	VII
Indian Bar Review	3	VIII
Journal of Criminal Justice	3	VIII
Emory Law Journal	2	IX
International Criminal Police Review	2	IX
Yale Law Journal	2	IX
CBI Bulletin	1	X
AME. Jurisprudence	1	X
George Washington Law Review	1	X
Asiatic Review	1	X
Uty of Chicago Law Review	1	X
Syracause Law Review	1	X
University of Tasmania Review	1	X
Delhi Law Review	1	X
Michigan Law Review	1	X
Col. Law Review	1	X
Jaipur Law Review	1	X
Journal of social history	1	X
Journal of Public Law	1	X
Popular Jurist	1	X
Journal of police sc & Amn.	1	X
Journal of criminal Law and criminology	1	X
Law Review	1	X

The above table clearly shows that 'All India reporter' is the most productive journal out of the 33 journals referred. 13 were cited not less than 3 times as given in the rank list. Only three journals cited twice and the rest (18 journals) were cited only once.

8.2 Type of documents cited in Law

In Law also the scholars prefer to use books as their major source of information (45.71%) and periodicals as their second preference source.

Table-40 Type of documents cited by Lawyers

Type of source	No. of citations	Percentage
Books	277	45.71
Periodicals	151	24.92
Govt. publications	140	23.10
Reports	30	4.95
Conference and seminars Procee.	- -	- -
Theses	- -	- -
News papers	8	1.32

8.3 *Authorwise distributions of citations in Law*

In Law also, it is seen that largest number of contributions are made by single authors (Table 41).

Table-41 Authorship pattern in Law

Type of authorship	No. of citations	Percentage
Single authors	407	85.50
Two authors	30	6.30
More than two authors	39	8.20

8.4 *Country-wise distribution of citations in Law*

The data given in table 42 reveals that scholars in Law cited more Indian publication (66.67%). In respect of cited articles USA rank second position.

Table-42 Country-wise distribution of citations in Law

Name of the country	No. of citations	Percentage
India	404	66.67
USA	94	15.51
UK	44	7.26
USSR	6	0.99
Others	58	9.57

9. **General Trends in the Distribution of citations in Social Science**

A consolidated table of citation pattern among social scientists in general is given in the following table and observations.

9.1 Age-wise distribution of citations in Social Science Literature

A consolidated statement of citations is also made here in order to understand year-wise distributions. (Table number 43) It can be seen that irrespective of the various types of social scientists, the majority of them prefer to cite literature published during 1970's (30.67%).

Table-43 Age-wise distribution of citations in social science literature

Year of publication	No. of citations	Percentage
Before 1950	979	15.64
1950 - 1959	854	13.64
1960 - 1969	1488	23.77
1970 - 1979	1920	30.67
1980 - 1989	1019	16.28

9.2 Core Journals in Social Science

Periodicals have assumed an increasingly dominant position in all the disciplines of social science. Some periodicals like 'Economic and political weekly' 'Journal of social psychology', 'Journal of Educational Psychology', 'India Today' etc. are used by researchers in more than one disciplines in social science.

9.3 Distribution of citations in social science among different forms of sources

It is seen that books (49.92%) are the major sources of citation, periodical literature (33.12%) followed by other sources viz govt. publications, reports, seminar/conference proceedings, theses, newspapers etc. with the percentage of 7.11, 3.80, 1.85, 2.92 and 1.28 respectively.

Table-44 Types of cited documents Vs social scientists

Type of source	Total citations	Percentage
Books	3125	49.92
Periodicals	2073	33.12

Govt. Publications	445	7.11
Reports	238	3.80
Seminars and conference Procee.	116	1.85
Theses	183	2.92
News papers	80	1.28

The above table shows the high percentage of citations of books and periodicals compared with other sources, tend to indicate that books and periodicals are most preferred channels of information. The important reasons may be the following. (i) periodicals and books are more easily accessible than conference proceedings and theses which many cases, are unpublished and therefore difficult to access. (ii) Articles in periodicals are published quickly and contain nascent thought useful to research scholars. (iii) periodicals and books are better organised bibliographically than theses, reports, conference proceedings etc.

9.4 Distribution by type of authorship according to subject

9.41 Single authorship

Table-45 Single authorship in social science literature

Subject	Total citations	Single authorship	Percentage
Psychology	1069	736	70.63
Education	1213	806	73.40
Geography	490	227	60.37
History	784	520	92.86
Political Science	586	399	85.99
Economics	724	478	82.41
Sociology	788	580	86.44
Law	606	407	85.50

In the case of social science, it appears that there is considerable number of single authorship in the subject like Psychology (70.63%), Education (73.40%), Geography (60.37%), History (92.86%), Political Science (85.99%), Economics (82.41%), Sociology (86.44%), and in Law (85.50%). The rate is very high in History and relatively low in

Geography.

9.42 Double authorship

The data given in table 46 below shows that double authorship is relatively high in Geography whereas it is low in History subject.

Table-46 Double authorship in social science literature

Subjects	Total citation	Double authorship	Percentage
Psychology	1069	224	21.50
Education	1213	209	19.05
Geography	490	103	27.39
History	784	34	6.07
Political Science	586	56	12.07
Economics	724	88	15.17
Sociology	788	73	10.88
Law	606	30	6.30

9.43 More than two authors

Table-47 More than two authors in social science literature

Subject	Total citations	More than two authors	Percentage
Psychology	1069	82	7.87
Education	1213	83	7.55
Geography	490	46	12.24
History	784	6	1.07
Political Science	586	9	1.94
Economics	724	14	2.42
Sociology	788	18	2.68
Law	606	39	8.20

It can be observed from the above table that among the research scholars in social science, scholars in Geography contribute relatively large number of publications (12.24%) by more than two authors. Literature contributed by more than two authors is relatively low in

History (1.07%).

9.5 Distribution of citations in Social Science by type of authorship

The following table shows the overall distribution of citations in social science in general by type of authorship.

Table-48 Authorship pattern in social science

Type of authorship	Total citations	Percentage
Single authors	4153	78.85
Two authors	817	15.51
Morethan two authors	297	5.64

In the case of social science, it is observed that in general single authorship is most cited. They have contributed 78.85% of the total contributions. The contribution made by the group of two authors are 15.50%. The third position goes to the group of more than two authors, which represents 5.64% of the total authors.

9.6 Country-wise distribution of citations

Table-49 Country-wise distribution of citations in Social Science

Name of the country	No. of citations	Percentage
India	2662	42.46
USA	1760	28.07
UK	888	14.16
USSR	90	1.44
Others	870	13.87

The consolidated data given in table 49 above shows that Indian publicaiions are most cited (42.46%) United States Publications has ranked next to India (28.07%). The UK has third position (14.16%). The analysis indicates that social scientists in general use Indian publications more than any other publications emanated from other countries. The publications from USA constitute their second preferred item.

10. Conclusion

It is hoped that the study, based on the analysis of bibliographic references appended to the theses, will provide an apparent view of information needs and use pattern of social scientist. It will also provide guidelines in the decision making process with regard to the acquisition and organization of documents in the social science libraries.

REFERENCES:

1. Martin, John: Citation Analysis. Journal of documentation 3(4); 1976 p. 290.
2. Britain, J. M: Information and its users. Bath, Bath University, Press, 1970. p. 129.

BIBLIOGRAPHY

Abdulsalam, C.A. Where does Kerala stand among Indian States and in the world : a statistical study based on the composite indices of Socio-economic development. In international Congress of Kerala studies. vol. 5. Thiruvananthapuram : AKG Centre for Research and Studies. 1994

Adeniran, O.R. Bibliometrics of Computer Science Literature in Nigeria. International Library Review. 1988; 20: pp. 347-359

Aiyepeku, W.O. The Bradford distribution theory: the compounding of Bradford periodical literature in geography. Journal of Documentation. 1977; 33(3): 210-219

Ajiferuke, I, etal. Collaborative Coefficient: a single measure of the degree of collaboration in research. Scientometrics. 1988; 14: pp. 421-433.

Alabi, G.A. The citation pattern of Nigerian Scientists. International library Review 1989; 21(1): pp. 129-137

Abafiatayo, Benjamin D. The research literature of agricultural economics. International library Review. 1989; 21: pp. 463-479

Al-Kharafi ,F, *et al* Science Research in Kuwait : a bibliometric study. Journal of Information Science. 1987; 13(1) : pp. 37-34

Arjunlal. Ranking of periodicals in the field of social sciences. Annals of library science and documentation. 1990; 37(2) : pp. 67-73

Arunachalam, S. Why is Indian Science mediocre. Science Today, 1990; Feb: pp. 8-9

Arunachalam, S and Hirannian, S. Journal preference of Indian researchers in. physical and chemical sciences. Proceedings of the seminar on Primary Communication in Science and Technology in India. 1978; 4-8 Dec: pp. 141-148

Arunachalam, S and Singh, U.N. Where do Israeli Scientists publish? Journal of Information Science. 1988; 14: pp. 221-231

Arunachalam, S. *et al* International collaboration in Science : Participation by the Asian Giants Scientometrics. 1994; 30(1): pp. 7-22.

Ashok Jain and Garg. K.C. LASER research in India: Scientific study and model projection. Scientometrics. 1992; 23(3): pp. 395-415.

Asundi, A.Y. and Humayoon Kabir, S. Evolving criteria for identification and selection of core periodicals in a subject: a case study of Indian Horticulture. Library Science with a slant to documentation and information studies.

Baker, D. Recent trends in the growth of chemical literature. Chemical Engineering News. 1976; 54: pp. 23-27

Baburajan, A. IEEE Transactions on computers : a bibliometric analysis. IASLIC Bulletin. 1988; 33 (2-3): pp. 53-59

Baird, L.M. and Oppenhein, C. Do citations matter? Journal of Information Science. 1994; 20(1): pp. 2-15

Balachandra Rao. Kerala's contribution to Mathematics and Astronomy. International Congress on Kerala studies, vol. 3. Thiruvananthapuram: AKG Centre for research and studies, 1994

Balasubramanian, S and Bhaskar, P.A. Citation analysis in neurology from the point of view of Indian Scientists. IASLIC Bulletin. 1984; 29: p. 63-67

Banik, B.D. and Bandhopadhyaya, S. Recent growth of periodicals on glass and ceramics-their rank. IASLIC-Bulletin. 1983; 30(1): pp. 23-30

Barker, Dale Lockard. Characteristics of the scientific literature cited by chemists of the Soviet Union. Dissertation: University of Illinois, 1966

Barre, R. Do not look for Scapegoats : link bibliometrics to social sciences and address societal needs. Scientometrics, 1994; 30[2]: pp. 419-424

Bartlet, M.S. The square root transformation in the analysis of variance. Journal of Royal Statistical society suppl. 1936; 3: pp. 68

Basak, N.D. and Das, B.B. Scatter of Indian Mathematician's contributions in primary periodicals. IASLIC Bulletin 1983; 5: pp. 67-73

Bavakutty, M and Mohamed Taheer. Flow of Islamic information through journal articles: a bibliometric analysis. Lucknow Librarian. 1991; 22(3-4): pp. 75-80

Beall, C. The transformation data from entomological field experiments so that the analysis of variance become applicable. Biometrika. 1945; 32: pp. 243-62

Begum, K.J. and Rejendran, N. Research collaboration in Zoological Science. IASLIC Bulletin. 1980; 35(2).

Begum, K.J. and Sani, L.K. Research collaboration in agricultural science. International Library Review. 1988; 20

Bek, M.T. Editorial statement. Scientometrics. 1978; 1: p.i.

Bernal, J.D. Preliminary analysis of pilot questionnaire on the use of scientific literature. Report of the Scientific Information Conference of the Royal Society London. Paper No. 46. pp. 589-561

Biog, F.S. and Loftman, K.A. Domestic and foreign periodicals in the field of petroleum chemistry: a statistical analysis. Oil and Gas Journal. 1949; 47: pp. 199-208.

Bird, P.R. Same sampling characteristics of bibliometric distributions. Journal of Information. 1977; 1(2): pp. 64-80

Blake, V.L.P. Faculty productivity, Journal prestige and school library media faculty. School library quarterly. 1994; 22(3): pp. 153-158

Bliss, N.J. International relationship : a bibliometric analysis of the field. Library and Information Science Research. 1994; 16(2): pp. 177-80.

Bookstein, A. Scientometrics : new opportunities. Scientometrics. 1994; 30(2-3): pp. 455-60.

Bookstein, Abraham. Bibliometric distributions. Library quarterly. 1976; 46(4): pp. 416-23

Bookstein, Abraham. Explanations of the bibliometric laws. Collection Management. 1979; 3: pp. 151-162.

Bookstein, Abraham. Patterns of Scientific Productivity and Social change : a discussion of Lotka's law and bibliometric symmetry. JASIS. 1977; 28(4): pp. 206-210

Booth, Andrew D. A 'Law' of occurrence for words of low frequency. Information and Control. 1967; 10: pp. 386-93

Bradford, S.C. Documentation. London: Crossly Lockwood, 1948

Bradford, S.C. Sources of Information on specific subjects. Engineering. 1934; 137: pp. 85-86

Braun, T, Glanzet, W and Schcebut, A. The bibliometric assessment of UK scientific performance. Scientometrics. 1991; 20: pp. 359-362

British Standards Institution, British Standards of Documentation Terms. London: BSI, 1976

Brittain, J.M. Information needs and application of the results of the user studies. Debons and cameson, Ed. Perspectives in information science, 1994

Broadus, R.N. Definition of bibliometrics: Scientometrics 1987; 12: pp. 373-379

Broadus, Robert N. The application of citation analysis to collection building. Advances in librarianship Vol. 7. New York: Academic Press, 1977.

Brookes, B.C. Bradford's law and the bibliography of science. Nature. 1969; 224: pp. 953-6.

Brookes, B.C. Numeric methods of bibliographic analysis. Library trends. 1973; 22(1): pp. 18-43

Brookes, B.C. Obsolescence of special library periodicals: Sampling errors and utility contours. JASIS. 1970; 21: pp. 320-9.

Brookes, B.C. The complete Bradford-Zipf bibliograph. Journal of Documentation. 1969; 25(1): pp. 58-60.

Brookes, B.C. The derivation and application of the Bradford-Zipf distribution. Journal of Documentation. 1968; 24(4): pp. 247-65.

Brookes, B.C. The foundation of Information Science. Part II: quantitative aspects: classes of things and the challenge of human individuality. Journal of Information Science. 1980; 2(209): pp. 21.

Brookes, B.C. Theory of Bradford's Law. Journal of documentation. 1977; 33(3): pp. 180-209.

Buckland, M.K. Are obsolescence and Scattering related. Journal of Documentation. 1972; 28(3): pp. 242.

Buckland, M.K. Book availability and the library user. New York: Pergamon Press, 1975

Bulick, Stephen. Book use as a Bradford-Zipf Phenomenon. College and Research Libraries. 1978; pp. 215-9.

Burnell, Q and Rousseau, R. Fractional counts for authorship attribution:

a numerical study. JASIS. 1995; 46(2): pp. 97-102.

Burton, Hilary D. Use of a virtual information system for bibliometric analysis. Information Processing and Management. 1988; 24(1): pp. 39-44.

Burton, R.E. and Keblee, R.W. The half life of same scientific and technical literatures. American Documentation. 1960; 11(1): pp. 18-22.

Cesare, R. D. The evaluation of G. C. impact in physical science using bibliometric indicators : Preliminary results. Proceedings of the Ist International Conference on Grey literature held at the RAI Congress Centre, Amsterdam, The Netherlands, 13-15 Dec. 1993, compiled by D. J. Frace. Amsterdam : Trans Atlantic, 1994.

Chakraborty, R. Citation analysis of Petrological literature, Lucknow Librarian. 1988; 15: pp. 99-104.

Chen, C. C. The use patterns of physics journals in a large academic research library. JASIS. 1972; 23(4): pp. 254-270.

Chen, Y. S. and Leemkuhler, F. F. A relationship between Lotka's law, Bradford's law and Zipf's law. Journal of the American Society of Information Science. 1986; 37: pp. 307-314.

Chen, Y. S., Chong, P. P. and Tong, M. Y. Dynamic behaviour of Bradford's law. JASIS - 1996; 46(5); pp. 370-383.

Chernyi, A. I. Distributions of the Zipfian type : an attempt at content interpretation. International Forum on information and documentation 1994; 19[1]]: pp. 3-7.

Chung, Y. K. Core international journals of classification systems : an application of Bradford's law. Knowledge organisation. 1994; 21(2): pp. 75-83.

Chung, Young M. A non-Bradfordian model for scattering of articles in a subject among periodicals. PhD. Thesis, USA: Case Western Reserve University, 1977

Clark, K. K. Americ's psychologists: A survey of growing profession. Washington: American psychologists Association, 1957.

Cline, Gloria Stark. A bibliometric study of two selected journals in library science. 1940-1974. Ph. D. Dissertation: University of Southern California.

Coile, Russel, C, Lotka's frequency distribution of Scientific productivity, JASIS 1977; 28(6): pp. 366-370.

Cole, F. J. and Eales, N. B. The history of comparative anatomy Part I : a statistical analysis of literature. Science Progress. 1917; 11: pp. 578-596.

Coleman, S. R. Disciplinary variables that affect the shape of Bardford's bibliograph. Scientometrics. 1993; 29[1]: pp. 59-81.

Cole, F. J. and Eales, N. B. History of comparative anatomy Part II, Statistical analysis of literature. Science progress. 1917; 11: pp. 578-596.

Cole, P. F. Analysis of reference questions records as a guide to the information requirements of scientists. Journal of Documentation. 1958; 14: pp. 197-207.

Coleman, A.M., Dhillon, D. and Coulthrad, B. A Bibliometric evaluation of the research performance of British University Politics Department: Publications in leading journals. Scientometrics. 1995; 32(1): pp. 49-66.

Conrad, G.M. Growth of biological literature and future of biological abstracts. Fed. Proc. 1957; 16: pp. 711-715.

Craig, G.R. Communication in Geology. Scottish journal of Geology. 1969; 5: pp. 305-321.

Crane, Diana. Transnational networks in basic sciences international organisations. 1971; 25(3): pp. 585-601.

CSIR Regional Research Laboratory: Annual Report -1994. Thiruvananthapuram: RRL (CISR), 1995.

Damodharan, V.K. Development of S & T infrastructure in Kerala. a data base instruction report. Thiruvananthapuram: STEC, 1991 (unpublished)

Daniel, R.S. Bibliometrics and Scholarly impact. American Psychologist. 1979; 34(8): pp. 725-728.

Dennis, W. Bibliography of eminent scientists. Science 1959; 79: pp. 180-183.

Dobrov, G.H. and Korennoi, A.A. The informational basis of scientometrics: on theoretical problems of informetrics. Hague :

FID 435, 1969.

Dondhue, Joseph. C. Understanding scientific literature: a bibliometric approach. Cambridge : MIT Press, 1973.

Drott, M.C. Reexamining the role of conference papers in scholarly communication. JASIS. 1995; 46(4): pp. 299-305.

Edwords, S. Bibliographic Instruction Research : an analysis of the journal literature from 1977 to 1991. Research strategies. 1994; 12(2): pp. 68-78.

Egghe, L. Little Science, big Science and beyond. Scientometrics. 1994; 30(2-3): pp. 389-92.

Egghe. L. Special features of the author-publication relationship and new explanation of Lotka's Law based on convolution theory. JASIS. 1994; 45(6): pp. 433-7.

Egghe, L. A theory of continues rates and applications to the theory of growth and obsolescence rates. Information processing and managements. 1994; pp. 30(2): pp. 279-92.

Egghe, L. An exact calculation of price's Law for Law of Lotka. Scientometrics. 1987; 11: pp. 81-97.

Egghe, L. Consequences of Lotka's law for the Law of Bradford. Journal of Documentation. 1985; 41: pp. 173-191.

Egghe, L. On the classification of the classical bibliometric laws. Journal of Documentation. 1988; 44: pp. 55-62.

Egghe, L. On the 80/20 rule. Scientometrics. 1986; 10: pp. 55-68.

Egghe, L and Rousseau, R. Editors. Informetrics 87/88. Select proceedings of I International Conference on Bibliometrics and Theoretical Aspects of Information Retrieval, Diepenbeck, Belgeum, 1987. Amsterdam: elsavier, 1988.

Egghe, L and Rousseau, R. Informetrics 89/90. Selection of papers submitted for II International Conference on Bibliometrics, Scientometrics and Informetrics, Ontario, Canada, 1989. Amsterdam: Elsavier, 1990.

Egghe, L and Rousseau, R. Introduction to informetrics. Amsterdam: Elsavier, 1990.

Elisha Raju, E. and Kavitha. Core journals in synthetic organic chemistry.

Annals of library science and documentation. 1985; 32(3-4): pp. 115-121.

Fairthorne, R.A. Empirical hyperbolic distributions (Bradford-Zipf-Mandelbrot) for bibliometric description and production. Journal of Documentation. 1969; 25(4): pp. 319-43.

Fang, P.H. and Fang, J.M. A modification of Lotka's function for scientific productivity information processing and management. 1995; 31(1): pp. 133-7.

Fano, R.M. Bibliographic Coupling. In Documentation in Action. New York: Reenhold; 1956.

Fano, R.H. Documentation in Action. New York: Reen-hold, 1956.

Fedorowicz, J. The Theoretical foundation of Zipf's law and its application to the bibliographic data base environment. Journal of American Society for information science. 1982, 33: pp. 285-293.

Fellee, W. An introduction to probability theory and its application. Ed. 3. (Vol. 1) New York: John Wiley and Sons, 1978.

Feller, W. Introduction to probability theory and its applications. Ed. 2(Vol.2) New York: Johnwiley and Sons, 1979

Fennee, P. and Lloyd, J. Serials important to geologists. Geotenes. 1970; 15: pp. 14-19.

Ferrante, Barbara Kopelock. Bibliometrics: access in the library literature. Collection Management. 1978; 2(3): pp. 199-204.

Finney, B. The reference characteristics of scientific texts. M.Sc. Dissertation, London. Centre for Information Science, City University, 1979.

Fox, Mary, F and Favu, Catherine, A. Independence and corporation in research, Journal of Higher Education. 1984; 55: pp. 347-59.

Frost, C. The use of citations in library research: Preliminary classification of citation functions. Library quarterly. 1979; 49(4): pp. 399-414.

Fusseler, Herman H. Characteristics of the Research Literature used by chemists and physicists in the United States. Dissertation. University of Chicago, 1948.

Fusseler, Herman H. Characteristics of the Research Literature used by chemists and physicists in the United States. Library quarterly. 1949; 19: pp. 19-36.

Fusseler, Herman H and Simon, Julian C. Patterns in the use of books in large research libraries. Chicago: University of Chicago, 1969.

Gapen, D.K. and Milner, S.P. Obsolescence. Library Trends. 1981; 30 (Summer): pp. 116

Garfield, Eugene. Citation analysis as a tool in Journal evaluation. Science. 1972; 178(4060): pp. 471-479.

Garfield, Eugene Citation frequency as a measure of research activity and performance. Current contents. 1977; 5(31): pp. 5-7.

Garfield, Eugene. Citation indexing - its theory and application in science, technology and humanities. New York : John Wiley, 1979.

Garfield, Eugene Essays of an Information Scientist. forward by Joshwa Ledeberg. Philadelphia: ISI Press, Vol. I. 1962-1973, Vol.2. 1974-1976.

Garfield, Eugene Highly cited articles XXVI: Some classic papers of the late 19th and early 20th century. Current. 1976; 21: pp. 5-9.

Garfield, Eugene. The mystery of transposed journal lists. Current contents (life Sciences). 1971; 14(31): pp. 5-6.

Garfield, Eugene and Sher, I.H. New factors in the evaluation of scientific literature through citation indexing. American Documentation. 1963; 14(4): pp. 195-201.

Garg, K.C. and Dutt, B. Bibliometrics of Agricultural research in India. Quarterly Bulletin of the International Association of Agricultural Information Specialists. 1992; 37(3): pp. 133-137.

Garrison, H.H. and others. Measuring characteristics of scientific Research: a comparison of bibliographic and survey data. Scientometrics. 1992; 24(2): pp. 359-370.

Garudadwajan, B., *et al*. Use of library documents in specified fields in NAL library. Library Science with a slant to Documentation 1981; 18(4): paper Q.

Garg, K. C. and Lalitha Sharma. A study of R & D indicators in Indian Industry using Bradford's law. Journal of Scientific and Industrial Research. 1994; 53 (Feb): pp. 81-87.

Glanzel, W and Schoepflin, U. A stochofic model for the ageing of scientific literature. Scientometrics. 1994; 30(1): pp. 49-64.

Glanzel, W and Schoepflin, U. Little Scientometrics, big Scientometrics and beyond Scientometrics. 1994; 30(2-3): pp. 375-84.

Goffman, William and Morris, Thomas. G. Bradford's law and library acquisitions. Nature. 1964; 226: pp. 922-923.

Goffman, William and Newin, Vaun A. Generalization of epidemic theory: an application to the transmission of leas. Nature. 1965; 224: pp. 225-8.

Goffman, W and Warren, K.S. Distribution of papers among journals based on mathematical analysis of two diverse medical literature. Nature. 1969; 221. pp. 1205-1207.

Goffman, W and Warren, K.S. Scientific information systems and the principle of selectivity. New York: Prager Press, 1980.

Gosnell, Chas F. Obsolescence of Books in college libraries. College and research libraries. 1944; 5: pp. 115-125.

Gosnell, Chas F. The rate of obsolescence in college library Book collections as determined by the Analysis of three select list of Books for college Libraries. Dissertation: New York University, 1943.

Griffith, Belvee C *et al*. Informal contracts in science: a probablistic model for communication process. Science. 1971; 173: pp. 164-5.

Gross, O.V. Bradford's law and the Keenan - Atherton data. American Documentation. 1967; 18: pp. 46.

Gross, P.L.K. and Gross, E.M. College libraries and chemical education. Science. 1927; 66: pp. 385-389.

Grupp. H and Hinze, S. International orientation, efficiency of and regard for research in East and West Germany: a bibliometric investigation of aspects of technology genesis in the United Germany. Scientometrics. 1994; 29(1): pp. 83-113.

Guha, B. Documentations and information services techniques and systems. Ed. 2. Calcutta: World Press, 1984.

Gundu Rao. Citation of documents in radio engineering: a case study.

DRTC Seminar 9. 1971; Paper B.E.

Gupta, A.K. Characteristics of documents cited by the Indian physicists: a case study. 1964; DRTC Seminar, 7. paper 13.

Gupta, B.M. and Nagpal, M.P.K. Citation analysis and its application: a review. Herald of library science. 1979; 18(1-2): pp. 88.

Gupta, D.K. Application of Bradford's law of citation data of Ethiopian Medical Journal. Annals of Library Science and Documentation. 1991; 38(3): pp. 183-188.

Gupta, D.K. Collaborative trends in early research of computer based storage and retrieval of geoscience information. IASLIC Bulletin, 1986; 31: pp. 55-64.

Gupta, D.K. Lotka's law and its application to author productivity distribution of psychological literature of Africa-Part I : for period 1996-1977 and Part II : 1966-1975. Herald of Library Science. 1989; 28: pp. 171.

Gupta, D.K. Lotka's law and productivity patterns of entomological literature in Nigeria for the period of 1900-1973. Scientometrics. 1987; 12: pp. 33.

Gupta, D.K. Scientometric study of chemical literature of Nigeria 1970-1984: Application of Lotks's law and 80/20 rule. Scientometrics. 1989; 15: pp. 171.

Gupta, D.K. and Gupta, S. Periodical literature of petroleum geology: a citation analysis study do determine obsolescence factors and patterns. IASLIC Bulletin. 1983; 28(2): pp. 75-86.

Harau, J.D. *et al*. Accumulative of advantage and disadvantage in research groups. Scientometrics. 1994; 29(I): pp. 239-51.

Hadagalt, P.B. Frequently cited periodicals by Indian agricultural economists: a citation analysis, IASLIC Bulletin. 1983; 28: p. 59-66.

Haiqi, Z. A bibliometric study on medicine, Chinese traditional (SIC) in Medicine database. Scientometrics. 1994; 31(3) pp. 241-250.

Hall, D.H. Rate of growth of literature in geoscience from computerised data bases. Scientometrics 1989; 17: pp. 15-38.

Hardie, B.G. and Voight, M.J. Use of periodicals in petroleum research.

Oil and Gas Journal. 1950; 49: pp. 121-6.

Hawkes, H.E. Literature of Geology. Geoteries. 1966; 10: pp. 23-24.

Hawkins, D.T. Unconventional uses of on-line information retrieval systems: on-line bibliometric studies. Journal of American Society for Information Science. 1977; 28(1): pp. 13-18.

Henkle, H.H. Periodical literature of biochemistry. Bulletin of Medical Library Association. 1988; 27: pp. 139-147.

Hersh, A.H. Drosophilla and course of research. Ohio Journal of Science. 1942; 49: pp. 121-6.

Hernou, P. Information life cycle: its place in the management of US Govt. Information sources. Government Information quarterly. 1994; 11[2]: pp. 143-70.

Herubel, J-P. V.M. and Buchanan, A.L. Citation studies in the humanities and social sciences : a Selefine and annotated bibliography.

Herubel, J-P. V.M. and Goedeken, E.A. Journals publishing American Library history : a research note. Libraries and culture. 1994; 29(2): pp. 205-9.

Heetzel, D.H. Bibliometrics, History of development of ideas in Encyclopaedia of Library and Information Science. New York: Marcel Dekker, 1985.

Hill, B.M. Zipf's law and prior distribution for the composition of a population. Journal of American Statistical Association, 1970; 65. pp. 1220-1232.

Hindle, Anthony and Buckland, Michael K. In library book usage in relation to circulation. Collection Management. 1978; 2(4): pp. 265-77.

Hirst, Grame. Discipline impact factors: A method for determining core journal lists. Journal of the American Society for Information Science. 1978; 29(4): pp. 171-172.

Hjreppe, R. A bibliography of bibliometrics and citation indexing and analysis. Stockholm, Sweden: Royal literature of Technology Library, 1980.

Hjreppe, R. Supplement to a Bibliography of bibliometrics and citation indexing. Scientometrics. 1982; 4: pp. 241-273.

Hodges, T. Citation indexing: its potential for bibliographic control. Ph.D. Thesis. Berkeley: University of California, 1978

Hooydonk, C.V. *et al.* A bibliotheconomic analysis of the impact factors of scientific disciplines. Scientometrics. 1994; 30(1): pp. 65-81.

Hubbert, J. J. Bibliometric models for journal productivity. Social Indicators Research. 1977; 4: pp. 44-73.

Hubert, John J. On the Naranan interpretation of Bradford's law. Journal of the American Society of Information Science. .1978; 27: pp. 339-41.

Hubbert, J.J. Relationship between the forms of Bradford's law. JASIS. 1978; 29(3): pp. 159-61.

Hulmes, E.W. Statistical bibliography in relation to the growth of modern civilization. London: Grafton, 1923.

Humayoon Kabir, S. Citations in bibliometrics research: an analysis. Library Science with a slant to Documentation and Information Studies 1993; 30(4): pp. 155-159.

Humayoon Kabir, S. Citations in Indian agriculture : an analysis. Annals of Library Science and Documentation 1990; 37(3): pp. 102-109.

Indian Science Abstract. 1994; 30(1). New Delhi: INSDOC, 1994

Jain, A.K. Sample data study of book usage in the purdue University Libraries. College and Research Libraries 1966; 27: pp. 13-8.

Jain, A.K. Statistical study of book usage. (Doctoral thesis). School of industrial Engineering, Purdue University, Fafayette, Indiana.

Kabir, S.H. Authorship trend and Solo research in bibliometrics: a bibiiometric study. Library Science with a slant to Documentation. 1994; 31(2): pp. 87-90.

Karisidappa, C.R. *et al.* Authorship pattern and collaborative research in psychology. IASLIC Bulletin. 1990; 35(2): pp. 73-78.

Karisiddappa, C.R., *et al.* Edi. Current studies in Library and Information Science. New Delhi: Manak Publications Ltd., 1993.

Ken, M.C. Bradford law and the journal titles cited by research scholars in Zoology at Ibadan University, Nigeria. IASLIC Bulletin. 1989; 34(3): pp. 97-104.

Kendall, M.G. Bibliography of operational research. Operational Research quarterly. 1960; 11(1-2): pp. 31-6.

Kent, Allen *et al*. Use of Library materials : the University of Pittsburgh Study. New York: Marcel Dekker, 1979.

Kesseler, M. M. Bibliographie coupling between Scientific Papers. American Documentation. 1963; 14(1): pp. 10-25.

Kochen, M. How well do we acknowledge intellectual depts? Journal of Documentation 1987; M: pp. 31-36.

Koening, M. and Harrl, T. Lotka's Law, Price's law and electronic publishing. JASIS. 1995; 46(5): pp. 386-388.

Kohut, J.J. A comparative analysis of obsolescence pattern of the U.S. geoscience literature. JASIS. 1974; Jul-Aug: pp. 242-251.

Kopelock, B. Bibliometrics: Access in the library literature. Collection Management. 1978; 2(3): pp. 199-204.

Kraukopf, M. Epistemometia, a term contributing to express the meaning and potential methodologies of scientometrics in Spanish speaking countries. Scientometrics 1994; 30[2-3]: pp. 425-8.

Kretschmer, H. Co-authorship networks of invisible colleges and institutionlised communities. Scientometrics. 1994; 30[1]: pp. 363-9.

Kretschmer, H. Quantity and quality in 'Science of Science'. Scientometrics. 1994; 30[2-3]: pp. 533-7.

Kuch, T.D.C. Relation of the title length to number of authors in Journal articles: In a brief communication in the Journal of American Society for Information Science. JASIS. 1978; 29: pp. 200-2.

Kuldip chand, Ed. Current trends in information technology: its impact on Information Science in India. New Delhi: Batra, 1993.

Kurian, C.T. Kerala's development experience: random comments about the past and some considerations for the future. International Congress of Kerala studies, Abstract, Vol. 1. Thiruvananthapuram: AKG Centre for research and studies, 1994.

Lalitha Kumari and Sengupta, I.N. Bibliometric evaluation of Biochemical research in India. Part II. Indian Journal of Information, Library and Society. 1990; 3(1-2): pp. 16-28.

Lancaster, F.W. Bibliometrics methods in assessing productivity and impact of research. Bangalore: Sarada Ranganathan Endowment for Library Science, 1991.

Lancaster, F.W. If you want to evaluate your library 2nd ed. London: Library Association, 1993.

Lancaster, F.W. Measurement and evaluation of library service. Washington: Information Resource Press, 1977.

Lawani, S.M. Bradford's law and the literature of agriculture. International library review. 1973; 5: pp. 341-350.

Lawani, S.M. Quality, Collaboration and citations in cancer research: a bibliometric study. (Ph.D. Dissertation), Florida State-University, 1980.

Lawani, S.M. and Serika, T.A.B. Some characteristics of the world literature on rice. International Rice Commission Newsletter. 1974; 23(1) p. 1-5.

Lazarev, V. S. Notion of a document : a centre of "gravity attraction" for getting metricians together. Scientometrics. 1994; 30[22-3]: pp. 511-16.

Lazorick, Gerald, J. Patterns of book use using the negative binomal distribution. Library Research. 1979; 1: pp. 171-188.

Leimkuler, F.F. An extract formulation of Bradford's law. Journal of Documentation 1980; 36(4): pp. 285.

Leinkuhler, F.F. Bradford distribution. Journal of Documentation. 1967; 23(3) pp. 197-207.

Leimkuhler, F. System analysis in University libraries. College and Research Libraries: 1966; 27: pp. 13-8.

Liang, L. and Mu, Y. Frequency distribution of Chinas' Scientific papers in terms of ranks. Journal of Library and Information Science. 1994; 20[2]: pp. 13-92.

Line, Maurice, B. Practical interpretation of citations and library use studies. College and Research Libraries. 1975; Sep: pp. 393-396.

Line, Maurice, B. The Half-life of periodical literature: apparent and real obsolescence. Journal of Documentation 1970; 26(1): pp. 46-52.

Line, M.B. and Sandison, A. Obsolescence and changes in the use of literature with time. Journal of Documentation. 1974; 30(3): pp. 283-350.

Lipetz, B. A. Improvement of the selectivity of citation indexes to science literature through the inclusion of citation relationship indicators. American Documentation. 1965; 16(2): pp. 81-90.

Lopez, S.J. Characteristics of Cassava Literature. Cassava Newsletter. 1980. 8. pp. 10-15.

Lotka, A. J. The Frequency distribution Scientific productivity. Journal of the Washington Academy of Science 1926; 16: pp. 317-323.

Looze, M-A. D. The application of Scientometric tools to the analysis of a sector in Plant biotechnologies : nitrogen fixation. Scientometrics. 1994; 30(1): pp. 23-24.

Lui, Z. Sensitivity to foreign countries : a comparison of US library literature about the US, 1980-1989. Libraries and culture. 1994; 29(2): pp. 210-19.

Luukkonen, T. Are we longing for the golden era lost or far the one to come. Scientometrics, 1994; 30(2-3): pp. 481-5.

Lynch, M.F. Variety generation a reinterpretation of shannon's mathematical theory of communication and its applications for information science. Journal of American Society for Information science. 1977; pp. 19-25.

Mahapatra, M. On the validity of the theory of exponential growth of scientific literature. Proceedings of 15th IASLIC Con. Bangalore, 1985: pp. 61-70.

Maheswarappa, B.S. Characteristics of phytopathology literature: a bibliometric study based on primary phyto pathology periodicals (1965-85). (Ph.D. Thesis), Gulberge University, Dept. of Library Science, 1992.

Maheswarappa, B.S. Research collaboration in different disciplines of applied science in India (1965-83): a bibliometric study. IASLIC Bulletin. 1987; 32(3): pp. 105-114.

Maheswarappa, B.S. and Nagappa, B. Indian phytopathology literature: a bibliometric study based on review of plant pathology. Journal of Library and Information Science. 1984; 9(1): pp. 36-47.

Maheswarappa, B.S. and Ningoji, M.M. Growth of literature in the field of applied science in India. (1965-1989). International Information Communication and Education. 1993; 12(2): pp. 191-200.

Maheswarappa, B.S. and Ningoji, N.M. Growth of literature in the field of S & T in India. International Information, Communication and Education (INICAE), 1992; 11(2): pp. 359-370.

Maheswarappa, B.S. and Prakash, B.P. Literature use pattern by researchers in the field of Botany. Journal of Library and Information Science. 1982; 7(1): pp. 15-32.

Maheswarappa, B.S. and Savadatti, S.G. Authorship pattern and collaborative research in plant breeding (1934-1989). IASLIC Bulletin. 1990; 35(3): pp. 119-124.

Mandelbort. An information theory of statistical structure of languages: Proceedings of symposium on application of communication theory. London: Butterworths, 1953.

Martens, B. and Saretzki, T. Quantitative analysis of thematic structures in the field of biotechnology: a study on the basis of conference data. Scientometrics. 1994; 30[1]: pp. 117-28.

Marticianem E. Shannon's entropy as a measure of the 'life' of the literature of a discipline. Scientometrics. 1994; 30(1): pp. 129-45.

Martin, B.R. British Science in the 1980s- has the relative decline continued. Scientometrics. 1994; 29(1): pp. 27-56.

Martyn, J. Bibliographie coupling. Journal of Documentation. 1964; 20: pp. 136-142.

Martin, John. Citation analysis. Journal of Documentation 1975; 31(4): pp. 290-297.

Martyn, J and Gilchrist, A. An evaluation of British Scientific periodicals. London: ASLIB, 1968.

Mathai, A.M. and Rathie, P.N. Basic concepts in information theory and statistics: Axiomatic foundations and applications. New Delhi: Wiley Eastern Ltd., 1975.

May, K. O. Quantitative growth of mathematical literature. Science. 1966; 154: pp. 1672-1673.

Mc Allister, P.R., *et al*. Comparison of peer and citation assessment of the influence of Scientific journals. JASIS. 1980; May: pp. 147-154.

Mc Grath, W.E. Correlating the subjects of books taken out of and books used within an open-stock library. College and Research libraries. 1971; 280: pp. 5.

Mc Grath, W.E. Little, Scientometrics, big scientometrics and beyond. Scientometrics. 1994; 30[2-3]: pp. 439-42.

Mc Grath, W.E. Inter-library loan analysis in collection development : a comment. Library Research and Technical Services. 1976; 20(1): pp. 101-3.

Mc Grath, William, E. Predicting book circulation by subject in a university library. Collection management. 1976-77; 1(3-4): pp. 7-25.

Mc Grath, W.E. Relationship between hard/soft, pure/applied and life/ non life desciplines and subject book use in a university. Information processing and Management. 1978; 14: pp. 17-28.

Mc Grath, W.E. Significance of books used as classified profile of academic departments. College and Research libraries. 1972; 33: pp. 212-9.

Meadows, A.J. Communication in Science. London: Butterworths, 1974

Meadows, A.J. Little Scientometrics, big scientometrics and beyond. Scientometrics. 1994; pp. 447-449.

Mehrotra, R and Lancaster, F.W. Where Indian Scientists publish. Current Science. 1984; pp. 684-687.

Mehard, H.W. Science: Growth and change. Massachussets: Hard Ward University Press, 1971.

Merton, R.K. and Garfield, E. Forward to little science, big science and beyond by D J de Solla Price. New York: Columbia University Press, 1963.

Miquel, F.J. Little scientometrics, big scientometrics and beyond. Scientometrics. 1994; pp. 443-5.

Miquel, F.J. and Okubo, Y. Structure of international collaboration in Science: Part 1. Comparisons of profiles in countries using a link indication. Scientometrics. 1994; 29[2]: pp. 271-97.

Montgomery, Leon, K *et al*. Cost-benefit model of library acquisition in terms of use: Progress report. JASIS. 1976; 27: pp. 73-4.

Morales, M. Informetrics and its importance. International forum for Information and Documentation. 1985; 10(2): pp. 15-21.

Moravisik, M.J. and Murugessan, P. Some results on the function and quality of citations. Social studies of science. 1975; 5: pp. 86-92.

Morse, Philip M. Library effectiveness: a system approach. Cambridge: MIT Press, 1968.

Morse, Philip M. Measures of library effectiveness. Library quarterly. 1972; 42(1): pp. 15-30.

Munshi, V.M., *et al*. Indian medical journals and Science citation indexes. IASLIC Bulletin 1990; 35(1): pp. 43-46.

Murphy, Larry J. Lotka's law in the humanities. In a brief communication in the journal of the American Society for Information Science. 1973; 24(6): pp. 461-2.

Nag, D.K. *et al*. Patterns of use of earth science literature by Indian geologists: a citation analysis proceedings of 15th IASLIC Conference. Bangalore: pp. 101-108.

Nagpal, P.S. Contribution of Indian universities to the Mainstream scientific literature: a bibliometric assessment. Scientometrics. 1995; 32(1): pp. 11-36.

Naranan, S. Bradford's law of bibliography of Science: an interpretation. Nature. 1970; 227: pp. 631-2.

Naranan, S. Power law relations in science. A self consistent interpretation. Journal of Documentation. 1972; 29(2) pp. 83-97.

Narendra Kumar, P. An international bibliometrics survey of periodical articles on bibliographical control. MLISc dissertation. 1994. Bangalore University.

Narendra Kumar. Literature contribution in journal of Environmental Biology: a bibliometric study. Annals of library science and documentation 1988; 33(1-2): pp. 31-40.

Narin, F. Patent bibliometrics. Scientometrics. 1994; pp. 147-55.

Narin, F. Evaluative bibliometrics: the use of publication and citation analysis in the evaluation of scientific activity. New Jersey: Cherry Hill, 1976.

Narin, F and Moll, Joy K. Bibliometrics: annual review of information science and technology. Vol. 12. New York: Knowledge industry publications, 1977.

Nederhof, A.J. and Meiju, R.F. Development of bibliometric indicators for utility of research to users in society: measurement of external knowledge transfer via publications in trade journals. Scientometrics. 1995; 32(1): pp. 37-48.

Neelameghan, A. Librametry. DRTC Seminar, Bangalore. 7. Paper L.C. 1969.

Neelameghan, A, *et al*. Discovery and duplication of antibiotics a correlation. Current Science. 1970; 34: pp. 511-2.

Neelameghan, A, *et al*. Distribution pattern of the duplication of discoveries. Nature. 1970; pp. 0-1060.

Neelameghan, A , *et al*. Universe of subjects and duplication of discovery: a case study using statistical methods. DRTC Seminar 7. Bangalore : DRTC, 1969.

New, Dorris, E and Ott, Retha Zune. Inter Library loan analysis as a collection development tool. Library Resource and Technical Services. 1974; 18(3): pp. 275-83.

Nicholas, D and Ritchie, M. Literature on Bibliometrics. London: Bingley, 1978.

Nijayunappa, R and Nijayunappa, P. Core journals used by Indian goescientists. (1978-82). Journal of geological society of India. 1985; 26: pp. 10-15.

Noor, Martyvonne, M. A quantitative analysis of the research articles published in core library journals of 1980. Library and Information Science Research. 1985; 7: pp. 261-273.

Novak, B. Use of books and journals in a large library for technological research. International Library Review. 1977; 9(2): pp. 241-6.

O'Connor, J.G. Growth of multiple authorship. DRTC Seminar 7. Paper JA. Bangalore: DRTC, 1969.

O'Connor, D and Voos, H. Emperical laws, theory construction and bibliometrics. Library Trends. 1981; 30: pp. 9-20.

Olsen, W.C. The current core literature of the agricultural sciences. Journal of agricultural and food information. 1994; 2[3]: pp. 99-109.

O'Neill, E.T. Effect of demand level on the optional size of journal collections. Collection Management. 1978; 2(3): pp. 205-16.

O'Neill, E.T. Generalised Bradford distribution. Research memorandum school of Information and library studies. Buffalo: State University of New York, 1972.

O'Neill, E.T. Limitations of the Bradford's distributions. Proceedings ASIS annual meetings. 10. 1973: pp. 177-8.

O'Neill E.T. Stochastic scattering model. Proceedings of ASIS.

Oppenheim, C and Renn, S.P. Highly cited old papers and the reasons why they continue to be cited. Journal of American Society for Information Science. 1978; 29(5): pp. 225-231.

Outturn of scientific and technical manpower in India, Vol. 1. New Delhi: HRD Group, CSIR, 1993.

Pair, C.I. Puberty or mid-life crisis. Scientometrics. 1994; 30[2-3]: pp. 517-20.

Pal, B.W. Economic survey of Aquiculture. Bombay: Kitab Mahal, 1961.

Pao, M.L. Collaboration in computational musicology. JASIS. 1982; 33: pp. 38-43.

Parvathamma, N. Bibliometrics- its origin and applications. In current studies in Library and Information Science. Ed. by C.R. Karisidappa and others. Delhi: Manak Publications, 1993: pp. 163-181.

Parvathamma, N. Trends in the growth of Indian earth science literature (1978-88): a bibliometric study.(Ph.D. Thesis). Gulberga: Gulberga University, 1990.

Parvathamma, N., *et al*. Core journals in earth science: a comparative study. Journal of Geological Society of India. 1991; 38: pp. 387-395.

Parvathamma, N, *et al*. Growth pattern of literature and scientific productivity of authors in Indian earth science (1978-88): a bibliometric study. Paper presented at 3rd International Con. on Informetrics. Bangalore, 9-11 Aug, 1991.

Pawan Sikka. Performance of scientific manpower in India. University news. 1994; 9 May.

Peritz, B. Little scientometrics, big scientometrics and beyond. Scientometrics. 1994; 30(2-3: pp. 414-16.

Peters, H. P. F. and Ŗaan, A. F. J. V. A bibliometric profile of top - scientometrics. 1994; 29(1): pp. 115-136

Pettigrew, K. E. and Nicholls, P. T. Publication patterns of LIS faculty from 1982-1982 : effects of doctoral programs, Library and Information Science Research. 1994; 16[2]: pp. 139-56

Peritz, B. C. A classification of literature roles for the social science and related fields. Scientometrics, 1983; 5(6): pp. 303-312.

Pichappan, P. Indian Agricultural Science Journals. IASLIC Bulletin. 1990; 35(2): pp. 83-86.

Pierret, J-D *et al*. Bibliometric analysis applied to Scientific review coverage of various aspects of sudden infant death syndrome. Documentaliste. 1994; 319[4-5]: pp. 205-10.

Portal, S. G. and Thompson, A. C. CUIB-METRIC : an integral system for metric analysis of bibliographic information. Information Bibliotechologica. 1994; 8[6]: pp. 27-31.

Potter, W. G. Introduction to Bibliometrics. Library Trends 1981; 30: pp. 5-7.

Pouris. A scientometric assessment of Agricultural research in South Africa. Scientometrics. 1989; 17(5-6): pp. 401-413.

Price, Derek J de solla. A general theory of Bibliometrics and other cumulative advantage process. Journal of American Society for information Science. 1976; 27(5): pp. 292-306

Price, D J de Solla. Little Science, big Science. New York: Columbia University Press, 1963.

Price, Derek. J de solla. Networks of Scientific papers. Science. 1965; 149: pp. 510-515

Price, D J de Solla. Some remarks on elitism in information and the invisible college phenomenon in science. JASIS. 1971; 22: pp.74-75.

Pritchard, Alan. Computers, statistical bibliography and abstracting services. (Unpublished), 1968.

Pritchard, Alan. Statistical Bibliography: An interim bibliography. London: New Western Polytechnic School of Librarianship, 1969

Pritchard, Alan. Statistical biblioghraphy or bibliometrics. Journal of Documentation. 1988; 25: pp. 179-191

Pritchard, A and Witting, G, Bibliometrics: a bibliography and index (1874-1959). Vol.1. Wartfard, UK: AAIM Books, 1960

Proceedings of the 15th IASLIC Conference on Bibliometric studies and current information. Bangalore, IASLIC, Dec. 1985

Qin, J. An investigation of research collaboration in the sciences through the philosophical transactions 1901-1991. Scientometics. 1994; 29(2): pp. 219-38.

Radhakrishnan, T and Kernizan, R. Lokta's law and computer science literature. JASIS 1979; 30: p. 51-54.

Ranganathan, S. R. Librametry and its scope. D.R.T.C Seminar 7 D. A. Bangalore: DRTC, 1969: pp. 285-301

Rangarajan, K. S. and Gupta, B. M. Analysis of journals for publications by Indian physicists. Journal of Library and Information Science. 1979; 4(2): pp. 144-161

Ranga Rao, B. V. Scientific research in India: analysis of publications. Journal of Scientific and Industrial Research. 1969; 26: pp. 166-176

Rann, A. F. J. V. Little scientometrics, big scientometrics and beyond. Scietometrics. 1994; 30[2-3]: pp. 529-31.

Raising, L Miles. Statistical Bibliography in the Health Sciences. Bulletin of the Medical Library Association. 1962; 50: pp.450-461.

Ravichandra Rao, I. K. Dispersion on survey analysis: Bradford and Pareto distributions. Library Science with a slant to Documentation. 1972. Paper T.

Ravichandra Rao, I. K. Distribution of Scientific productivity and social change. JASIS. 1980; 31(2): pp. 111-121.

Revichandra Rao, I. K. Distribution of transactions over users and over documents in Canadian academic libraries: (a dessertation proposal). School of Library and Information Science; The University of Western outario, London

Ravichandra Rao, I. K. Document and user distribution in transaction records of Canadian University Libraries. (Doctoral thesis). London: School of Library and Information Science, University of Western Onterio, 1981.

Ravichandra Rao, I. K. Document and user distribution in Transaction records of Canadian University Libraries. (Doctoral Theses). Ontraio, Canada: University of Western Ontario, 1981

Ravichandra Rao, I. K. Entropy of a probability distribution of transactions/users: a measure of concentration of transaction/users were documents. Library Science with a slant to Documentation, 1979. Paper G.

Ravichandra Rao, I. K. Obsolescence and utility factors of periodical publications. DRTC Seminar 9. Paper J.

Ravichandra Rao, I. K. Quantitative methods for library and Information Science. New Delhi, Wilay Eastern, 1983.

Revichandra Rao, I. K. Workshop on Scientific communication and bibliometrics. Calcutta, 1988.

Ravichandra Rao. I. K. and Meera, B. M. Growth and obsolescence of literature: an empirical study. Paper presented at 3rd International Conference on Informetrics. Bangalore, 9-12 August 1991.

Reguant, S and Casadella, J. English as a Linga Franga in geological Scientific Publications : a bibliometric study. Scientometrics. 1994; 29[3]: pp. 335-51

Rouse, W. B. Tutorial : Mathematical modelling of library systems. JASIS 1989; 30: pp. 181-192

Rousseau, R. Similarities between informetrics and econometrics. Scientometrics. 1994; 30[2-3]: pp. 385-7.

Russell. J. M. Back to the far informetrics. Scientometrics. 1994; 30(23): pp. 407-10.

Sandison, A. Use of older literature and its obsolescence. Journal of

Documentation. 27(3): pp. 184.

Sangam, S. L. Citation analysis of doctoral dissertations in social sciences accepted by the Karnataka University, Dharward during 1964-1982. (Ph. D. Thesis). Gulbarga: Glbarga University, 1985.

Sangam, S. L. Obsolescence of literature in economics. ILA Bulletin: 1989: 25(2): pp. 59-66.

Sanjay Mishra, IASLIC Bulletin: a bibliometric study. IASLIC Bulletin. 1990; 35(3): pp. 125-128.

Saurer, W and Weinberager, R. Planetary nebule : Some statistics on a continuously growing field and its contributions. Scientometrics. 1994; 31[1]: pp. 85-95.

Scales, Pauline A. Citation analysis as indicator of the use of serials: a comparison of ranked titles lists produced by citation counting and from use data. Journal of Documentation. 1976; 32(1): pp. 17-25.

Schorr, Alan Edward. Lotka's law and map librarianship. In the brief communication in the journal of the American Society for Information Science. 1975; 26(3): pp. 189-90.

Schrader, A. M. Teaching Bibliometrics. Library Trends 1981; 30(6): pp. 151

Schubert, A. Quantitative studies of science: a current bibliography. Scientometrics. 1983; 5: pp. 125-133, 397-403. 1984; 6: pp. 55-59, 197-202 and 359-367.

Scientist and Technologists in Kerala: a Directory. Thiruvananthapuram: STEC, Government of Kerala, 1989

Scott, J. T. Medical writing : Should you add to it? Learned Publishing. 1994; 1[1]: pp. 29-31

Sen, B. K. From bibliometrics to informetrics : Workshop on scientific communication and bibliometrics held at ICCB and INSDOC, Calcutta, Feb. 15-26, 1988.

Sen, B. K. and Chattergee, B. K. An introduction to researchers in Bibliometrics. Part 1. Back ground and Perspective. IASLIC Bulletin. 1990; 35(3): pp. 105-117

Sen, B. K. and Kumar, N. Indian contributions in bibliometrics. (1958-84): a review. Annals of Library Science and Documentation. 1986: 3: pp. 85-103

Sen, B. K. and Lakshmi, V. V. Bibliometrics in India 1988: a review. ILA Bulletin. 1991; 26(4): pp. 198-206

Sen, B. K. and Nerendra Kumar, Indian contribution in bibliometrics, 1958-1984: a review. Annals of Library Science and Documentation. 1986; 33(3): pp. 85-103.

Sen, B. K. and Shailendra, K. Evaluation of recent scientific research output by a bibliometric method. Scientometrics. 1992; 23(1): pp. 21-30

Sen, S. K. and Chatterjee, S. K. An introduction to researches in bibliometrics. Part 2. Glossary. IASLIC Bulletin. 1990; 35(4): pp. 155-170

Sengupta, I. N. Bibliometrics: a bird's eye view, IASLIC Bulletin. 1985; 30(1): pp. 168.

Sengupta, I. N. Bibliometrics and identification of core periodicals. Herald of library science 1990; 29(3-4): pp. 226-247

Sengupta, I. N. Bibliometrics and its applications. Information science and libraries ed. by Pushpa Dhyani. A.New Delhi: Atlantia, 1990.

Sengupta, I. N. Bibliometircs research: growth of biomedical literature Vol. i. Calcutta: S. B. A. Pub., 1988.

Sengupta, I. N. Choosing pharmacology periodicals: study of the growth of literature in the field. Annals of library science and documentation. 1974; 21: pp. 1-21.

Sengupta, I. N. Growth of bio-chemical literature. Nature. 1973; 244: pp. 75-76

Sengupta, I. N. Impact of Scientific serials in the advancement of Medical knowledge : an objective method of analysis. International Library Review. 1972; 4: pp. 169-195

Sengupta, I. N. Physiology periodicals. International library review. 1974; 6: pp. 147-165.

Sengupta, I. N. Recent growth of the literature in biochemistry and changes in ranking of periodicals. Journal of documentation. 1973; 29: pp. 192-211

Sengupta, I. N. A weightage formula to rerank bio-chemical periodicals. International Library Review. 1984; 16: pp. 44-458

Shailendra, K., *et al*. Bibliometrics : a tool for Scientific evaluation. ILA Bulletin 1989; 24(4): pp. 194-207.

Shannon, C. E. A mathematical theory of communication. Bell system Technology Journal. 1948; 27: 379-423; 623-56.

Shockly, William. On the statistics of individual variation of productivity in research laboratories. Proceedings of IPE. 279-90.

Shtrikman, S. Some comments on Zip's law for the chinese language. Journal of Information Science. 1994; 20[2]: pp. 142-3.

Simon, Herbert. A. On a class skew distribution functions. Biometrika. 1955; 42: pp. 425-40

Singh, R. S. Ranking of periodicals in chemistry from the point of view of Indian Scientists. Annals of library science and documentation. 1974; 21: pp. 55-67

Small, H. Co-citation in the Scientific literature : a new measure of the interrelationship between two documents. Journals of the American Society for Information Science. 1973; 24 (4): pp. 266-269

Spies, B. R. Effectiveness of journal in exploration geophysics. Geophysics. 1991; 56[6]: pp. 4-858.

STEC Review 1983-1994 [Annual Report]. Thiruvananthapuram: STEC, Govt. of Kerala, 1994.

S tieg, L. Technique for evaluating the college library book collection. Library quarterly. 1943; 13; pp. 33-44.

Subramanyam, K. Bibliometric studies of research collaboration. Journal of Information Science. 1983; 6[1]: pp. 333-38 Subramanyam, K. Core journals in computer science. IEEE Trans, Prof. Communication. 1976; PC 19: pp. 22-25

Subramanyam, K. Scattering, laws of Encyclopaedia of library and information Science. 1979; 26: pp. 336-354

Subramanyam, K and O' Pecko, M. M. Environmental research journals. Environmental Science and Technology. 1979. 13. pp. 327-929.

Subramanyam, K and Stephens, Eisie, M. Research collaboration and funding in Biochemistry and chemical engineering. International

Forum of Information and Documentation. 1982; 7: pp. 26.

Synder, H, Cronin, B. and Davenport, E. What's the use citation? Citation analysis as a literature topic in selected disciplines of the social sciences. Journal of information Science. 1995; 21(2): pp. 75-85

Szava - Kovats, E. : First fact - finding examination of a phenomenon of Scientific literature. Journal of Information Science. 1994; 20[1]: pp. 155-70.

Tagliacozzo. Citations and citations indexes. Match. Inform. Med. 1967; 6: pp. 136-142.

Tague, J. M. Success-breeds-success phenomenon and bibliometric process. Journal of American Society for Information Science. 1981; 32[4] pp. 250-6

Tague, J. *et al.* The law of exponential growth. Evidence, implications and forecasts, Library Trends. 1981; Summer: pp. 125-149

Theci, Henry. Economics of Information theory. Amsterdam. North Holland Publishing Co., 1969.

Thorne, F. C. The citation index : another case of spurious validity. Journal of clinical psychology. 1977; 33: pp. 1157-1161.

Thorpe, Peter, Inter-library's loan analysis for journal organization. Aslib proceedings. 1979; 31[7]: pp. 352-9.

Thyer, B. A. and others. Academic affiliation of social work journal article authors: a publication productivity analysis form 1984-1988. Journal of Social Service Research. 1994; 18[3-4]: pp. 153-167.

Tibbetts, P. The method for estimating the in-house of periodical collection in the University of Minnesota Biomedical Library. Bulletin of the Medical Library Association 1974; 62: pp. 37-48.

Trueswell, R. W. Two characteristics of circulation and their effect on the implementation of mechanised circulation control systems. College and Research Libraries. 1964; 25: pp. 285-91.

Trueswell, R. W. User circulation satisfaction Vs size holdings at three academic libraries. College and Research libraries 1969; 30: pp. 204-13.

Tuck, R. K. Introduction to the Principles of Agricultural Economics, Edinburgh: Neil and Co. Ltd., 1961

Turner, Stephen J. Identifier method of measuring use as applied to modelling the circulation use of books from a University Library. JASIS. 1977, 28(2): pp. 96-100

Turnover of science manpower declining. Indian Express Daily. 1993; Dec. 3: pp. 13.

Vasukuttan, K. A. Modern Science in Malayalam : a historical descriptive and critical study on technical terminology. Ph. D. Thesis, Thiruvananthapuram: Dept. of Malayalam, University of Kerala, 1991 (unpublished).

Vickery, B. C. Bradoford's law of Scattering Journal of Documentation. 1948; 4(3): pp. 198-203.

Vijayakumar, K. P. The legacy of Kerala in scientific research. Souvenir and Abstract of National symposium on environmental dimensions of palynological science. Thiruvananthapuram: ERRC & PSI, 1995 April.

Vilanilam, J. V. Science Communication and development. New Delhi : Malhotra Publishing house, 1993.

Vos, H. Lotka and Information Science. JASIS, 1974; 25: pp. 270-272

Wagher-Dobler, R. and Barg, J. The dependence of Lotka's Law on the selection of time periods in the development of Scientific areas and authors. Journal of Documentation. 1995; 51[1]: pp. 28-43.

Weinstock, Melvin. Citation indexes. Encyclopaedia of Library and Information Science. Vol.5. New York. Marcell Dekker, 1971: pp. 16-40

White. Emilie, C. Bibliometrics: Curiosity to convention. Special libraries. 1985 Winter : pp. 35-42.

Whitney, G. Pattern of authorship in major bibliographic data bases, Scientometrics. 1993; 26(2): p. 275-292.

Wilkinson, Elizabeth. Ambiguity of Bradford's law, Journal of Documentation. 1972; 28(2): pp. 122-130

Willams, C. B. Number of publication written by Biologists. Annals of Human Genetics. 1944; 12: pp. 145-6

Witting, Glenn R. Statistical bibliography - a historical footnotes. Journal of Documentation. 1978; 34(3): pp. 240-1

Workshop on scientific Communication and Bibliometrics. 15-26 Feb. 1988. Calcutta: IICB and INSDOC, 1988.

Woutcss, P. and Ley Clesdarib. Has price's dream come ture? Is scientometrics a hard science: Scientometrics. 31[2] 1994; 31[2]: pp. 193-222.

Yablonsky, A. I. On fundamental regularities of the distribution of scientific productivity. Scientometrics. 1980; 2[1]: pp. 3-34

Yitzhaki, M. Relation of title length of journal articles to number of author. Scientometrics 1994; 30[1]: pp, 321-32.

Yuasa, M. Centre of Scientific activity, its shift form the 16th to 20th century. Japanese studies in the history of science. 1962; 1: pp. 527

Zipf, George, Human behaviour and the principle of least effort, New York: Addison-Wesley Press, 1949.

Zipr, G. K. Human behaviour and the principle of least effort : an introduction to human ecology. Cambridge, Massachusetts : Addison- Wesiey. 1949.

Zuker Man. Nobel Laureates in Science, patterns of productivity, collaboration and authorship. Am. Social Review. 1967; 32: pp. 391-403.

Zunde, Pranas and Slamecka, Vladimer; Distribution of indexing terms for maximum efficiency of information transmission. American Documentation. 1967; 18[2]: pp. 104-108.

INDEX